EXPERTISE

PHILOSOPHICAL REFLECTIONS

Other interview books from Automatic Press ♦ $\frac{\vee}{\text{I}}$P

See all published and forthcoming books in the 5 Questions series at
www.vince-inc.com/automatic.html

EXPERTISE
PHILOSOPHICAL REFLECTIONS

Evan Selinger

Automatic Press ♦ ∨—P

Automatic Press ♦ $\frac{\vee}{\mathsf{I}}$P

Information on this title: www.vince-inc.com/automatic.html

© Automatic Press / VIP 2011

First published 2011

Printed in the United States of America
and the United Kingdom

ISBN-10   87-92130-37-2   paperback
ISBN-13   978-87-92130-37-2   paperback

Typeset in LaTeX2$_\varepsilon$
Cover design by Vincent F. Hendricks

Contents

Contents

Preface

Expertise: Philosophical Reflections
by Evan Selinger

◆

Problems related to expertise are amongst the most pressing issues of our time. Although philosophical reflection has much to contribute to the identification and resolution of these problems, the topic of expertise has not received the professional attention that it deserves. Of course, notable exceptions exist. For example, my co-edited volume *The Philosophy of Expertise* (Columbia University Press 2006) showcased exemplary instances of philosophical writing on the following topics: trusting experts, expertise and practical knowledge, and contesting expertise. *Expertise: Philosophical Reflections* is a sequel. As such, it covers both conceptual and applied issues, often collaboratively. While this style may be unusual for some forms of philosophical inquiry, it is appropriate and perhaps even necessary for analyzing topics as complex and multi-disciplinary as the ones treated here.

Section I, "Expertise and Phenomenology," critically addresses issues related to Hubert Dreyfus's phenomenological conception of expertise.

Chapter 1, "Dreyfus on Expertise: The Limits of Phenomenological Analysis," begins by clarifying Dreyfus's provocative thesis that expert authority across domains and fields is rooted in intuitive judgment. It then criticizes his phenomenological account of how expertise develops, explaining why the famous five-stage model misrepresents the relation between social influences and expert cognition. It additionally clarifies some of the practical implications that follow from this mistake as they relate to socially significant matters of trust.

Chapter 2, "Chess Playing Computers and Embodied Grandmasters: In What Way Does the Difference Matter?" critically examines Dreyfus's depiction of the fundamental differences between how human Grandmasters—who he characterizes as a paradigm case of intuitive experts—and machines play chess. Revisiting the social influence thesis, the case is made that Dreyfus establishes his demarcation of human and computer chess styles by unduly depicting Grandmaster perception as immune to the influences of eros and prejudice. This empirically inaccurate depiction is shown to be underwritten by a normative agenda, one that mars Dreyfus's descriptions of the expertise at issue. Specifically, Dreyfus aspires to preserve a putatively "endangered" vision of expertise from the threats he sees arising from educational and technological endeavors structured around the computational conception of mind. In allowing this type of activism to taint descriptive scholarship, I claim that Dreyfus weakens his claim to offer a robustly embodied philosophical account of expertise, and inadvertently depicts Grandmasters as more like machines than flesh-and-blood people.

Section II, "Interactional expertise," critically addresses issues related to Harry Collins's concept of "interactional expertise."

Chapter 3, "Interactional expertise and Embodiment," clarifies why Collins's main thesis about interactional expertise stipulates that one can acquire all of the linguistic understanding of a domain by immersing oneself in the language of the domain without actually engaging in the hands-on dimensions of its practices. Turning to phenomenology, I argue that, on the one hand, contrary to what Dreyfus maintains, Collins correctly identifies a genuine form of knowledge that is validated by empirical data acquired through his adaptation of the Turing Test. On the other hand, I contend that Collins overlooks important developmental considerations concerning how interactional expertise is acquired. In making concrete suggestions for what some of those considerations might entail, I suggest that the phenomenological conception of embodiment might explain more about the type of knowledge that interactional experts possess than is considered permissible under the constraints stipulated by Collins's concept of "linguistic socialization."

Chapter 4, "The Incompatibility of Industrial Expertise and Sustainability Science," argues that because "wicked problems" transcend the scope of normal, industrial-age science, problems such as sustainability require reform of current scientific practices. Crucially, universities should cultivate research and education programs devoted to "sustainability science," an enterprise that can be formally demarcated from "business as usual" and "systems optimization" approaches. Advancing sustainability science requires a shift in orientation away from reductionism and intellectual specialization towards integrative approaches to science, education, and technology that draws upon ethical awareness and which requires the cultivation of *interactional expertise* to facilitate cross-disciplinary exchange. Unfortunately, existing education and research training are ill-equipped to prepare scientists and engineers to operate effectively in a wicked problem milieu. It is therefore essential to create new programs of graduate education that will train scientists and engineers to become *sustainability science experts* equipped to recognize and grapple with the myriad of ethical and transdisciplinary dimensions embedded in their technical research programs.

Section III, "Ethics and Politics of Expertise," critically addresses a range of conceptual and normative issues related to expert authority.

Chapter 5, "Feyerabend's Democratic Argument Against Experts," revisits Paul Feyerabend's scathing portrayal of modern experts as a threat to democracy. While commentators have questioned the accuracy of Feyerabend's analysis of experts, they have not assessed the accuracy of his depiction of laypeople. On the basis of providing this assessment, I contend that although Feyerabend has political reasons for wanting to demythologize grandiose notions of expertise, his political project hinders clear thinking about the question by idealizing the alternative "lay perspective".

Chapter 6, "Catastrophe ethics and activist Speech: Reflections on Moral Norms, Advocacy, and Technical Judgment," critically examines whether there are ethical dimensions to the way that expertise, knowledge claims, and expressions of skepticism intersect on technical matters that influence public policy, especially during times of crisis. It compares two different perspectives on the matter: a philosophical outlook rooted in discourse and virtue ethics and a sociological outlook rooted in the so-called third wave approach to science studies. The comparison occurs through metaphilosophical analysis and applied claims that clarify how the disciplinary orientations appear lead to different judgments about matters related to Robert Paarlberg's condemnation of activists who advise African politicians to ban genetically modified food.

Chapter 7, "Competence and Trust in Choice Architecture," critically examines Richard Thaler and Cass Sunstein's theory of how designers can use "nudges" to improve decision-making in various situations where people have to make choices. The claim is advanced that the moral acceptability of nudges hinges in part on whether Thaler and Sunstein can provide an account of the competence required to offer nudges, an account that would serve to warrant our general trust in choice architects. What needs to be considered, on a methodological level, is whether Thaler and Sunstein have clarified the *competence required for choice architects* to prompt subtly our behaviour toward making choices that are in our best interest from our own perspectives. The arguement is made that, among other features, an account of the competence required to offer nudges would have to clarify why it is reasonable to expect that choice architects can understand the constraints imposed by "semantic variance". Semantic variance refers to the diverse perceptions of meaning, tied to differences in identity and context, that influence how users interpret nudges. The chapter concludes by suggesting that choice architects can grasp semantic variance if Thaler and Sunstein's approach to design is compatible with insights about meaning expressed in science and technology studies and the philosophy of technology. But this might require choice architects to *develop expertise in identifying meaning and making predictions on issues for which meaning matters.* Whether such expertise actually exists or can be cultivated, however, is a matter that goes beyond the scope of the present inquiry.

Acknowledgements

This book could not have been written without the help of the following collaborators: Harry Collins, Robert Crease, Tom Seager, Paul Thompson, and Kyle Powys Whyte. My deep appreciation extends to all of them. Noreen Selinger deserves special praise for being supportive during the trials and tribulations that come with academic professionalization. My thanks also go to Vincent Hendricks for his continued friendship and support of this project and his assistant editor Henrik Boensvang of Automatic Press ♦ $\frac{\vee}{\text{I}}$ P. Finally, I would like to acknowledge the assistance provided by the College of Liberal Arts at Rochester Institute of Technology. The Paul and Francena Miller Faculty Fellowship enabled this book to be completed.

The following articles are being re-printed here with the gracious permission of the respective presses:

Evan Selinger and Robert Crease, "Dreyfus on Expertise: The Limits of Phenomenological Analysis." *Continental Philosophy Review* 35 (2002): 245-279.

Evan Selinger, "Chess Playing Computers and Embodied Grandmasters: In What Way Does the Difference Matter?" In Ben Hale, ed., *Philosophy Looks at Chess.* Open Court Press (2008): 65-87.

Evan Selinger, "Feyerabend's Democratic Argument Against Experts." *Critical Review* 15, nos. 3-4 (2003): 359-373.

Evan Selinger, Harry Collins, and Hubert Dreyfus, "Interactional Expertise and Embodiment." *Studies in History and Philosophy of Science* 38 (2007): 722-740.

Evan Selinger, Harry Collins, and Paul Thompson, "Catastrophe ethics and activist Speech: Reflections on Moral Norms, Advocacy, and Technical Judgment." *Metaphilosophy* 42, nos.1-2: 118-144.

Evan Selinger and Kyle Whyte, "Competence and Trust in Choice Architecture." *Knowledge, Technology, Policy* 23, nos. 3-4 (2010): 461-482.

New York, January 2011
Evan Selinger

Acknowledgements

Part I
Expertise & Phenomenology

◆

1

Dreyfus on Expertise: The Limits of Phenomenological Analysis

Written with Robert Crease

Introduction

Expertise is of central importance to contemporary life, in which many economic, political, scientific, and technological decisions are routinely delegated to experts (Barbour, 1993, pp. 213-223). Citizens defer to the authority of experts not only in circumstances involving technical dimensions, but also in "all sorts of common decisions" (Walton, 1997, p. 24). On the one hand, routine deference to experts has political consequences and some scholars even suggest that it undermines rational democratic procedures and communicative action by allowing ideology to substitute for critical discussion (Turner, 2001).[1] On the other hand, volatile controversies over issues with a scientific- technical dimension can result in the suspension of routine deference and increased suspicion towards experts. It is hardly surprising, therefore, that the nature and the proper criteria for identifying expertise have been hotly debated in political and legal contexts. In legal contexts, for instance, the question of the proper criteria for expertise regularly arises in connection with developing the appropriate criteria for certifying expert witnesses.[2]

Expertise is an issue for which philosophical clarification seems

---

[1] Turner discusses the bases of Stanley Fish, Jurgen Habermas and Michel Foucault's concerns about why expertise undermines liberal democracy. He argues that these concerns "depend upon a kind of utopianism about the character of knowledge that social constructionism undermines" (Turner, 2001, p. 147, n. 7).

[2] The problem of certifying expert witnesses is frequently discussed in relation to the issue of "junk science" (See Black, Ayala and Saffran-Brinks, 1994; Jasanoff, 1992; Caudill and Redding, 2000; Huber, 1991).

appropriate and even essential. The 1993 landmark decision by the U.S. Supreme Court concerning the use of expert witnesses, *Daubert v. Merrell Dow Pharmaceuticals,* appealed to several concepts of philosophical origin, with particular attention given to Karl Popper's notion of "falsification" (Huber and Foster, 1999, pp. 3748). Philosophers themselves occasionally have been called upon to serve as expert legal witnesses while medical ethicists have advised hospital boards, politicians, and U.S. Presidents concerning such politically sensitive programs as the Human Genome Project and stem cell research (Nussbaum, 2001; Ruse, 1996).[3]

Aside from its social implications, the issue of expertise is also philosophically important for several reasons. One is that it bears on the philosophy of mind. The classical locus of expertise, or "expert knower," is the subject, and the way experts understand and are attuned to the world bears on the nature of subjectivity, intentionality, and rationalist and representational notions of consciousness (Pappas, 1994). The nature of expertise, for instance, is the focal point around which turns the debate over whether intelligence can be success- fully disembodied in artificial intelligence (AI) schemes, expert systems, and computer-based distance learning programs (Collins, 1995).

A second reason for the philosophical importance of the question of expertise is that it crystallizes the conflict between two traditions, classical philosophy of science and science and technology studies (henceforth STS). Classical philosophy of science takes expertise for granted and assumes the legitimacy of an expert-lay divide, while STS takes a skeptical approach towards experts for granted, presupposing the need to expose their illegitimacy (Mialet, 1999, pp. 552-553).

Finally, addressing expertise stimulates the interface between phenomenology and the sciences. Though some have argued that phenomenological description is only capable of capturing subjective dimensions of experience, and hence is inappropriate to use when trying to understand science (Latour, 1999, p. 9), scientific

---

[3] When theorists assume an expert social role based upon the prescriptive dimensions of their research then the problem of what Winner calls the "values expert" emerges (Winner, 1995, pp. 65-67). Winner argues that applied theorists not only provide counsel in "interminable" debates, but: tend to misconstrue the audiences they address as an overly generalized "we," fail to recognize how social change is instituted, and occasionally help legitimate already made political decisions.

practice requires the development, exercise, and coordination of a variety of expert skills that are open to phenomenological clarification.

Nevertheless, philosophers have rarely addressed the subject explicitly, though implicit and unexamined notions of expertise often lurk under rubrics such as "authority," "colonization," "power," and "rational debate" (Turner, 2001). Hubert Dreyfus is one of the few to have overtly addressed the concept, and this chapter is devoted to critically appraising his account. This analysis of Dreyfus will proceed in five steps. We shall: (1) place his model of expertise in an embodied context, (2) outline his general conception of skills, (3) summarize his descriptive model of expertise, (4) present his normative theory regarding which expectations about experts are justifiable, and (5) point out certain problems with his account. We argue that Dreyfus, by proposing that fundamental expert characteristics can be specified independently of cultural and historical considerations, demonstrates the importance of phenomenology to the subject by showing persuasively that expertise cannot be examined exhaustively by sociological, historical, and anthropological analyses. But we also identify certain descriptive and normative problems in Dreyfus' s account. While Dreyfus shows phenomenologically that experts cannot be reduced to ideologues and artifacts of social networking, he also lacks hermeneutical sensitivity by overstating the independence of the expert and expert decision- making from cultural embeddedness.

1. Expertise and the Body

The significance of Dreyfus's account of expertise can be highlighted by considering how two traditions, classical philosophy of science and Science and Technology Studies (STS), avoid addressing the issue. Each in effect treats creative expert performance as something extraordinary that 'just happens,' for which no satisfactory methodological weight can be given. The goal of traditional philosophy of science, for instance, is the rational reconstruction of the organizational dimensions that are at the root of science's efficacy and objectivity, with particular emphasis on how its operation is only temporarily disrupted by anomalies. It approaches creativity as a predominantly mental act for which true philosophical discussion can take place only about its products. Where creative ideas come from is not considered a proper epistemological question and is relegated to psychology or history within the framework of the distinction between context of discovery and context of justification (Mialet, 1999, 552). Contemporary STS, on the other hand treats expertise as "distributed," externalized into particular settings like the laboratory and social networks, standardized in technologies, criteria of scientificity, protocols for evaluating proof, and the rhetorical means of recruiting allies (Mialet, 1999, 552). To do otherwise, according to STS proponents, would risk "naturalizing" expertise, leading to undue authority being conferred on experts and to the repression of lay knowledge, values, and interests. By focusing on how experts become overly exalted through processes of mediation, STS proponents place a nondistributed sense of expertise into a "black box" (Mialet, 1999, 553).[4]

Even though traditional philosophy of science and STS have different reasons for wanting to demystify expertise, they produce similar results. For all their sharp disagreements, both traditional philosophy of science and STS refrain from discussing the relation between expertise and the body. They both agree that to demystify expertise and provide an accurate account of science, invariant features of bodily praxis need to be ignored.

The neglect of embodiment broaches the traditional phenomenological theme, reaching back to Husserl, that the practical involvement of living bodies ultimately grounds the knowledge that they have about the world, including abstract-scientific knowledge.[5] Because he shows how expert judgment is an instance

---

[4] "Black box" is an engineering term used in science studies analogously to the Marxist concept of hegemony. "Black boxing," like hegemony, refers to background assumptions that are generally regarded as self-evidently true and not requiring further investigation (Feenberg, 1995, 7). To open the "black box" of scientific expertise is to show, through close empirical and conceptual analysis, that what appears to be self-evidently true, culturally sanctioned, and not requiring further investigation about experts is false, hidden from cultural scrutiny, and in dire need of critical analysis. When the "black box" of expertise is opened, STS theorists contend that experts do not emerge as self-sufficient geniuses whose knowledge is infallible, certain, and objective. Scientific experts are revealed rather to be non-extraordinary, biased people whose successes and failures emerge from working within a competitive network of distributed knowledge and prestige. From the STS perspective, it is principally because the network's operation is rarely explicitly described and theoretically examined that nonexperts mistakenly perceive scientific experts as more knowledgeable, authoritative, and trustworthy than is appropriate.

[5] Abstract contexts alone do not allow scientists to mathematize, model, and formalize the world. Body oriented skills are used to operate the technological instruments that stabilize phenomena in order for scientists to ma-

of embodied human performance, Dreyfus's account of expertise has been quite influential. Like classical philosophers of science and STS proponents, he seeks to demystify expertise. The way he does so is to place expertise on a continuum with everyday lifeworld practices, rather than isolating it from them. This, of course, reflects the basic phenomenological tenet that all practical and theoretical activities, no matter how abstract their outcomes, need to be understood on a continuum with basic lifeworld practice. Experts, he insists, merely act the way each of us does when performing mundane tasks: "We are all experts at many tasks and our everyday coping skills function smoothly and transparently so as to free us to be aware of other aspects of our lives where we are not so skillful" (Dreyfus, 1990, 243).

Dreyfus's skill model is intended not only to be compatible with everyday experience, but also with Merleau Ponty's notion of the lived body (*le corps vécu*) and his concepts "intentional arc" and "maximal grip."[6] He intends to descriptively enhance the phenomenological approach, to "lay out more fully than Merleau Ponty does" how skills are acquired, improved, and used, by closely attending to "how our relation to the world is transformed as we acquire a skill" (Dreyfus, 1999a, 1). It is critical, for instance, to describe the *embodied, situated subject* noting,"under what conditions deliberation and choice appear," in order to avoid "making the typical philosophical mistake of reading the structure of deliberation and choice into [his/her] account of everyday coping" (Dreyfus, 1990, 239). Dreyfus also contends that his model of how experts act is empirically verified by evidence in neural

---

nipulate and interpret them. Ihde repeatedly argues that as with mundane uses of technology, the technological instruments used in scientific settings extend and transform bodily praxes through "embodiment relations"; they are absorbed and incorporated into bodily experience of the world like Heidegger's hammer or Merleau-Ponty's blind person's cane, and the phenomena that scientists can produce change as the forms of embodiment change (Ihde, 1998, 42-43).

[6] Dreyfus defines "intentional arc" and "maximal grip" as follows: "The *intentional arc* names the tight connection between the agent and world, viz. that, as the agent acquires skills, those skills are 'stored,' not as representations in the mind, but as dispositions to respond to the solicitations of the world. *Maximal grip* names the body's tendency to respond to these solicitations in such a way as to bring the current situation closer to the agent's sense of an optimal gestalt. Neither of these abilities requires mental or brain operations" (1998; 1999a).

networks, specifically in Walter Freeman's account of the brain dynamics underlying perception. (1999a; 1999b, 6-10; 1998)[7]

After situating expertise on a continuum with everyday skillful activity, and tying it to classical phenomenological concepts such as those provided by Merleau-Ponty, Dreyfus attempts to provide a list of characteristic skills and affective feelings that are essential for someone to possess in order to be considered a genuine expert. He lays the groundwork for what might be called a metaphysics of expertise (though he prefers the term "ontological" to 'metaphysical'). This is theoretically significant because the term 'expertise' is habitually articulated without sufficient critical attention being paid to similarities and differences in its use. STS studies suggest that expertise is a culturally dependent phenomenon, meaning that its definition changes in relation to the historical transformations that govern how it is perceived. In this sense, STS considers "expertise" to be a concept that fits Raymond Williams profile of a "key word," a term whose meaning is "inextricably bound up with the problems [it is] being used to discuss" (Williams, 1976, 13). By contrast, Dreyfus proposes that fundamental expert characteristics are general and invariant, which means they can be specified independently of cultural and historical specificity.

2. Expertise and Skills

Dreyfus first developed the basis for his descriptive account of expertise with his brother Stuart during the 1960's, when hired by the RAND corporation as a consultant to evaluate their work on

---

[7] Although we believe that Dreyfus turns to the theme of "brain topics" simply to establish another perspective that can be made compatible with his phenomenological approach, Sheets-Johnston argues that the reference to neural networks and brain functions contradicts his phenomenological aspirations: "[We] find that any erstwhile sense we might have had of a phenomenological subject has given way to a neurological one, while at the same time we find a phenomenological subject to be ostensibly fully present but transformed in ways utterly foreign to our experience in that it 'presumably senses' its own 'brain dynamics'" (2000, 357). Moreover, she argues that, "Not only can they [neural nets] not distinguish between formal and informal learning, but the very vocabulary by which they operate is not simulation-friendly to the latter kind of learning and is thus inappropriate to its description and deflective to its understanding" (Sheets-Johnstone, 2000, 357-358). I believe that what Sheets-Johnston forgets is that some still consider phenomenology to be a subjective form of analysis and that the reason why Dreyfus probably gestures towards neural nets is to demonstrate that phenomenology can reveal objective structures of bodily praxis.

artificial intelligence. His research culminated in a paper called "Alchemy and Artificial Intelligence" (Dreyfus, 1967) and the book *What Computers Can't Do* (Dreyfus, 1992).[8] In *Mind Over Machine: The Power of Human Intuition and Expertise in the Era of the Computer* the two brothers developed a model of expert skill acquisition whose scope was claimed to be universal (Dreyfus and Dreyfus, 1986). They aimed to provide a phenomenological account of how adults acquire skills by instruction in all fields involving skilled performance, whether of the intellectual or practical kind.[9]

It is no exaggeration to characterize this model of expert skill acquisition as a touchstone in Dreyfus's career. He uses it, for instance, as the basis to: (1) demythologize the hype associated with artificial intelligence projects, in particular, "expert computer systems" designed to simulate human expertise (Dreyfus and Dreyfus, 1986 and Dreyfus, 1992); (2) judge social biases that "endanger" professional experts (such as nurses, doctors, teachers, and scientists) by imposing "rationalization" constraints (Dreyfus and Dreyfus, 1986); (3) explain what is wrong with dominant tendencies in American styles of business management (Dreyfus and Dreyfus, 1986); (4) defend the accuracy of Merleau-Ponty's non-representational account of intentionality and action (Dreyfus, 1988); (5) expose the practical limits of Jürgen Habermas's neo-Kantian conception of ethics (Dreyfus, 1990); (6) explain the expertise of political action groups (Dreyfus, 1997); (7) clarify

---

[8] During the 1960s, when Dreyfus first formulated his critique of artificial intelligence and its failed hype, the intellectual atmosphere at the Artificial Intelligence Laboratory at the Massachusetts Institute of Technology was overly hostile to recognizing the implications of what he said, and as a result, he almost lost his job. By contrast, Winograd notes that today, "some of the work being done at that laboratory seems to have been affected by... Dreyfus" (1995, 110). Certainly Dreyfus has made some erroneous predictions: that a computer would never beat him at chess, that computers would never play master level chess, and that a computer would never be world champion of chess. Yet when critics emphasize these failures, they also pay him a compliment, since unimportant forecasts are not worth noting.

[9] According to Dreyfus, the primary desideratum of phenomenology has always been to adequately describe expertise (even if historical phenomenologists such as Husserl, Heidegger, and Merleau-Ponty did not use the term 'expertise' in their analyses), because at bottom, phenomenology aims at getting behind the prejudices that impede how human experience is understood (1986, 2-5).

what *techne* and *phronesis* mean for Martin Heidegger and Aristotle, and in doing so, correct his acclaimed, book-length interpretation of Heidegger (Dreyfus, 2000); (8) explain the relevance of Kierkegaard's normative stages of the subject's development to evaluations of the Internet's value as a medium for communication (Dreyfus, 2001); and (9) critique the educational potential of "distance learning" programs (Dreyfus, 2001).

Dreyfus focuses on two types of skill: intellectual and motor.[10] Intellectual skills provide the ability to use one's knowledge effectively and readily in the execution or performance of a task. Motor skills involve the coordination of dexterity in the execution of learned physical tasks. By characterizing expertise as the acquisition of either of these skills, Dreyfus suggests that expertise pertains to the augmentation of the subject's aptitudes, abilities, and powers. A first key element of Dreyfus's account is a rejection of the common tendency to define experts as sources of information. Expert skills are principally a matter of practical reasoning, of "knowing how" rather than "knowing that." The latter is propositional knowledge of and about things, obtainable through reflection and conscious appreciation. Dreyfus associates "know how," such as walking, talking, and driving, with practical knowledge that is mostly experienced as a "thoughtless mastery of the everyday" and does not require conscious deliberation for successful execution (Dreyfus, 1990, 244). In many instances, knowing how to do something such as riding a bicycle, recognizing a complex visual pattern, producing a coherent and grammatical sentence, and solving a physics problem involves the exercise of inarticulatable skills of which one cannot fully give an account. Taking a famous example from Michael Polanyi, Dreyfus writes:

> You probably know how to ride a bicycle. Does that mean you can formulate specific rules that would successfully teach someone else how to do it? No you don't. You can ride a bicycle because you possess something called "know how," which you acquired from practice and sometimes painful experience. The fact that you can't put what you have learned into words

---

[10] Sheets-Johnston argues that by analytically separating intellectual from bodily skills that Dreyfus assumes a "pernicious" "Cartesian split" between mind and body by forgetting how bodily skills are foundational for intellectual skills (2000, 355-356).

means that know-how is not accessible to you in the form of facts and rules. If it were, we could say that you "know that" certain rules produce proficient bicycle riding. (Dreyfus and Dreyfus, 1986, 16)[11]

It is possible to suggest hints and maxims that approximate elements of smooth performance, but seeing the point of these hints, and being capable of following these maxims, to a large extent presuppose the skill that the hints and maxims are supposed to account for. Moreover, Dreyfus holds that once skill is acquired, one tends not to follow the maxims used during the initial stages of learning.

Following Heidegger, Dreyfus argues that practical agents tend only to reflect on how their experience is organized during "breakdown" scenarios when pre-thematic styles of environmental coping prove insufficient for accomplishing ordinary goals. In other words, people only reflect upon how they do what they do when what they do fails to effectively work. The justification for this derivative use of reflection is also practical, based on the agent's experience that to reflect upon what one does and the rules for doing it usually leads to practical problems doing what one wants to do. Going from pre-conscious behavior to a conscious appreciation of the rules followed during particular actions marks the agent's transition from practical to theoretical reasoning, from "knowing how" to "knowing that." Connecting this abstract point with concrete experience, Dreyfus writes, "Have you ever been driving effortlessly along a city street in a stick shift car and suddenly started thinking about the gear you are in and whether it is appropriate? Chances are the sudden reflection upon what you were doing and the reasons for doing it was accompanied by a severe degradation of performance" (Dreyfus and Dreyfus, 1986,7). For Dreyfus, the most basic tasks requiring skilled performance are

---

[11] Although there are many surface similarities between Dreyfus and Polanyi on the issue of tacit knowledge, there also is a notable difference. Dreyfus argues that while Polanyi recognizes that formalisms cannot account for the tacit performance of riding a bicycle, he still believes that such performance is governed by "hidden rules": "The reference to hidden rules shows that Polanyi, like Plato, fails to distinguish between performance and competence, between explanation and understanding, between the rule one is following and the rule which can be used to describe what is happening" (1992, 330-331). Dreyfus also differs from Kuhn's analysis of tacit knowledge since this analysis primarily situates tacit dimensions of knowledge in the general structure of a paradigm (Stengers, 2000, 6).

best accomplished if one does not consciously focus on what one is doing.

Finally, Dreyfus contends that skills are flexible ranges of response and even physical skills cannot be reduced to a repetitive series of kinesthetic movements. He contends, "A skill, unlike a fixed response or set of responses can be brought to bear in an indefinite number of ways" (Dreyfus, 1992, 249). Joseph Rouse elaborates this point well in a discussion of Dreyfus. In learning to throw, he points out, what is involved is not a series of repetitive motions, "but a range of responses to throwable things. Learning to throw overhand means that one can also throw sidearm, though the movement is different. Having learned to imitate a fairly limited number of sentences, I can produce an unlimited variety of different ones" (Rouse, 1987, 61). Thus to learn a flexible range of response is not to memorize an "actual movement" or "thought pattern" with the intention of repeating them, but to grasp of a field of possibilities" (Rouse, 1987, 61)

3. The Descriptive Model

Following *Mind Over Machine*, Dreyfus's model of expertise has appeared nearly verbatim in a number of articles.[12] In these articles, he never challenges the core theses about, and descriptions of, expertise articulated in *Mind Over Machine*, but rather expands the range of what the model can be applied to, and comments on the epistemological and metaphysical features of expert performance that he previously applied but did not fully articulate.[13]

At the empirical level, Dreyfus bases his model of expertise on invariant patterns found in descriptions of skill acquisition relayed by first-person testimonies of "airplane pilots, chess players, automobile drivers, and adult learners of a second language" who discussed how they learned to make "unstructured" decisions

---

[12] Although I will in the later sections of the article be critical of Dreyfus's model of expertise, it would be wrong to mistake errors in the model for the development of something less than a model. This is exactly what Sheets-Johnston implicitly does by dwelling on Dreyfus's phrase "skill story," as if he were simply presenting a 'hypothetical' account (Sheets-Johnston, 2000, 355-357).

[13] The only time Dreyfus alters his five-stage model is when he adds two additional stages—mastery and wisdom—in his recent reflections on the Internet (2001).

(Dreyfus and Dreyfus, 1986, 20).[14] For Dreyfus, this empirical dimension is supposed to have *phenomenological justification*. Traditionally, the concept of justification relates propositions to a public sphere that can demonstratively verify or refute the content of, or logical or inferential connections in, statements. From a phenomenological perspective, propositions are justified when they detail experiential invariants that all subjects are capable of recognizing as according with their own experience. This is signaled in the following invitation Dreyfus extends to his readers: "You need not merely accept our word but should check to see if the process by which you yourself acquired various skills reveal a similar pattern" (Dreyfus and Dreyfus, 1986, 20). Thus, Dreyfus's account is not supposed to be vulnerable to accusations

---

[14] Dreyfus takes his target group involved in making "unstructured" decisionsto be paradigmatic of the "typical learner" and classifies "a common pattern" observable in their behavior (1986, 20). The adjective "typical" denotes a class of learners who "possesses innate ability" and also have "the opportunity to acquire sufficient experience" (Dreyfus, 1986, 20). "Unstructured" and "structured" are standard terms discussed in theories of information management. They are used to classify organizational differences in the range of decision making, usually with "semi-structured" appearing as a mid-point in this range. When Dreyfus references "structured" decisions, he means the type of decisions that are made when "the goal and what information is relevant are clear, the effects of the decisions are known, and verifiable solutions can be reasoned out" (Dreyfus, 1986, 20). In other words, "structured" decisions involve well-understood situations that have a common procedure for handling them. "Structured" decisions can be found in situations that are repetitive, routine, and have the pertinent pieces of evidence remaining stable as time passes. Examples of "structured" decision making can be found in situations where "context-free" deliberation dominates, such as, "mathematical manipulations, puzzles, and, in the real world, delivery truck routing and petroleum blending" (Dreyfus, 1986, 20). In contrast with "structured" decisions, Dreyfus characterizes "unstructured" ones as: intuitive, commonsensical, heuristic, and involving trial and error approaches. He states they have a tendency to be ad hoc, are not programmable, and "contain a potentially unlimited number of possibly relevant facts and features, and the ways those elements interrelate and determine other events is unclear" (Dreyfus, 1986, 20). The situations that require "unstructured" decisions typically are elusive ones: where it is not possible to specify in advance most of the decision procedures to follow; where the decision maker must provide judgment, evaluation, and insights into the problem definition whose parameters cannot be precisely identified; where unquantifiable factors are central; and where there is no agreed-upon procedure for making decisions. These types of decision are routinely made by people in "management, nursing, economic forecasting, teaching, and all social interactions" and require "considerable concrete experience with real situations (Dreyfus, 1986, 20).

of mischaracterizing idiosyncratic experts as paradigmatic or begging the question of how one knows what an exemplary expert field is. Because of adhering to the phenomenological method and its justificatory mechanisms, his account is also not meant to be susceptible to counterexamples that could disprove the universality of its scope. It is putatively immune to these types of criticism because his skill model is expected to be meaningful in such a way that all experts can rediscover and verify its essential elements for themselves.

At the theoretical level, his model is developmental and envisions skill acquisition as occurring through five ascending stages: (1) an initial "beginner" phase, to (2) an "advanced beginner" phase, then to (3) a "competent" phase, then to (4) a "proficient" phase, and (5) finally the educational process culminates when students achieve "expert" status. Dreyfus contends that the successful expert passes through all five stages and in the order described. Each stage is cumulative and predicated upon successfully accomplishing and going beyond what was developmentally possible in the previous stage. In the first stage, the beginner, who "wants to do a good job," learns a "context free" set of "rules for determining action" and tends to act slowly in remembering how to apply them (Dreyfus and Dreyfus, 1986, 21). In the advanced beginner stage, the student who now has more "practical experience in concrete situations" begins to "marginally" improve by recognizing "meaningful additional aspects" of the situation that are not codified by rules (Dreyfus and Dreyfus, 1986, 22-23).

Beginners and advanced beginners typically experience their commitment to a practice as "detached" while a competent performer feels "involved" in the outcome of his or her performance (Dreyfus and Dreyfus, 1986, 26). In the competent stage the learner frequently feels "overwhelmed," as if he or she is "on an emotional roller coaster," having to cope with "nerve-wracking and exhausting" aspects of the practice and feels "overloaded" due to facing too many potentially relevant elements to remember (Dreyfus, 2001, 35). Consequently, the competent learner narrows down those elements, devises a "plan" and chooses a "perspective" in order to selectively address "relevant features and aspects" of the situation (Dreyfus and Dreyfus, 1986, 26-27). By making these changes, the competent performer experiences "a kind of elation unknown to the beginner," including "pride" and "fright" (Dreyfus, 1985, 117-118). Accepting responsibility means "mistakes are felt in the pit of the stomach". (Dreyfus, 1985, 118)

In the proficient stage, the student transcends what Dreyfus calls the "Hamlet model of decision making," the "detached, deliberative, and sometimes agonizing selection of alternatives," which typifies the first three stages of skill acquisition (Dreyfus and Dreyfus, 1986, 28). Here the performer's reliance on rules and principles for seeing what goals need to be achieved is largely replaced by "know-how," an *"arational"* grasp of the situation that Dreyfus calls "intuitive behavior"; although the proficient performer must still contemplate and deliberate about what to do to achieve his or her goals (Dreyfus and Dreyfus, 1986, 27-36).[15] "Action becomes easier and less stressful" as the proficient performer "simply sees what needs to be done rather than using a calculative procedure to select one of several possible alternatives" (Dreyfus, 2001, 40).

In the final stage, the expert not only sees what needs to be done, but also how to achieve it without deliberation, immediately, yet "unconsciously" recognizing "new situations as similar to whole remembered ones" and intuiting "what to do without recourse to rules" (Dreyfus and Dreyfus, 1986, 35). Thus, the expert, like masters in the "long Zen tradition" or Luke Skywalker when responding to Obi Wan Kenobi's advice to "use the force," transcends "trying" or "efforting" and "just responds" (Dreyfus, 1999b, 22, fn.13). Dreyfus summarizes the "fluid performance" of expertise as: *"When things are proceeding normally, experts don't solve problems and don't make decisions; they do what normally works"* (Dreyfus and Dreyfus, 1986, 30-31). He even claims that at the expert level the ability to distinguish between subject and object disappears: "The expert driver becomes one with his car, and he experiences himself simply as driving, rather than as driving a car" (Dreyfus and Dreyfus, 1986, 30). When the expert experiences the "flow" of peak performance he or she "does not worry about the future and devise plans" (Dreyfus and Dreyfus, 1986, 30). By being immersed in the moment, the expert can

---

[15]Dreyfus writes: "Although irrational behavior... should generally be avoided, it does not follow that behaving rationally should be regarded as the ultimate goal. A vast area exists between irrational and rational that might be called *arational*. The word rational, deriving from the Latin word *ratio*, meaning to reckon or calculate, has come to be equivalent to calculative thought and so carries with it the connotation of 'combining component parts to obtain a whole'; arational behavior, then, refers to action without conscious analytic decomposition and recombination. *Competent performance is rational*; *proficiency is transitional; experts act arationally"* (1986, 36).

experience "euphoria," which athletes describe as playing "out of your head". (Dreyfus and Dreyfus, 1986, 40)

Whereas some STS researchers argue that it is untenable to decisively distinguish the expert from the nonexpert because their apparent differences are social illusions, Dreyfus contends that beginners and experts can be *demarcated* in three different ways.[16] A first demarcation is based on the expert's "immersion into experience and contextual sensitivity." The expert differs from the beginner because he or she no longer relates principally to a practice analytically through context-free features that are recognizable without experience. Instead, through skillful behavior rooted in experience, he or she recognizes important features as contextually sensitive (Dreyfus and Dreyfus, 1986, 35). This situational engagement leads to a change in the expert's judgment: "[The] novice and advanced beginner exercise no judgment... and those who are proficient or expert make judgments based upon their prior concrete experience in a manner that defies explanation" (Dreyfus and Dreyfus, 1986, 36). A second demarcation centers on the temporal connection between action and decision making. While slowly following rules and deliberating characterize the beginner's actions, the expert's actions are immediate and intuitive situational responses. A final point of demarcation is that affective transformation accompanies the development of expertise. In passing through developmental stages the expert's subjectivity and relation to the world are transformed in a manner that qualitatively differs, and can be demarcated from, a beginner's relation to the world. While in early stages the learner is "frustrated" and "overwhelmed," in the last stage the expert, who through learned that the outcome of a situation matters by making risky commitments, sheds those affects and enjoys "fluid" and "smooth" performance. In his critique of John Searle's account of the background and its relation to intentionality, Dreyfus describes an expert tennis player as being able to become so absorbed in the "flow" of the game that he or she no longer feels the pressure to win, and only responds to the gestalt tensions

---

[16] As Michael Callon puts this point: "Researchers in the wild participate in the subversion of modern institutional framing by challenging the oppositions that we had come to take for granted, yet that are crucial, such as the distinction between expert and layperson" (Callon, 2001).

on the court (1999b, 4-5).[17]   Additionally, the change in affect
from beginner to expert corresponds with a change in meaning.
To illustrate this point Dreyfus cites a quote from Bobby Fischer,
"perhaps history's greatest chess player": "'chess is life"'' (Dreyfus
and Dreyfus, 1986, 33).

Despite the differences detailed in these demarcations it is just
as important, from Dreyfus's perspective, to highlight the continuities. According to Dreyfus even the most abstract practice
maintains an essential, though sometimes hidden, relation to the
lifeworld. This is why he avoids using the terms 'nonexpert' and
'layperson,' which would antonymically suggest someone who did
not have the special skill or knowledge found in a particular field;
instead, he uses "beginner," suggesting someone on one end of a
spectrum of potentialities. Although not every practice provides
every beginner with an opportunity to achieve expert level mastery, many do: "Not all people achieve an expert level in their
skills. Some areas of skill have the characteristics that only a
very small fraction of beginners can ever master the domain of"
(Dreyfus and Dreyfus, 1986, 21).

It is also important to emphasize the *foundational implications*
of his model. Different fields are organized around different types
of essential skills; still, his account is a model of skill acquisition
that can be formally specified without reference to any particular

---

[17] This emphasis on affect is important for Dreyfus's criticism of so-called
"expert" computer systems, which he argues can approximate competent human performance. Skill acquisition for Dreyfus involves not only a cognitive
acquisition by the subject but also an affective transformation that computers
cannot experience. This point is especially significant in serving to highlight
Dreyfus's phenomenological background. Phenomenologists have long argued
that to be a subject is to have an intentional relation to the world such that
changes made to the subject correlate to changes made to the subject's world.
The subject who goes through the developmental process of expert apprenticeship is not the same subject as the one who began the process, and the
world is not meaningful for the subject in the same way; experts and nonexperts are indeed different subjects. They are different types of people who
deliberate and feel differently, and to whom the world responds differently.
Not only can experts do more things than beginners, but their whole affective demeanor changes. The way that experts care about their activities in
a practice changes from when they were beginners, progressing from relative
detachment to engaged commitment. This why Dreyfus characterizes the five
developmental stages as "qualitatively different perceptions" of what a task
is and what mode of decision-making is appropriate to handling that task
(1986, 19).

field. Of course diverse fields define experts in different ways due to the type of content that typifies the field. For example, a relation to winning defines an expert chess player, whereas a relation to arriving at a destination defines an expert driver. Nevertheless, a fundamental definition of expert as intuitive, committed, rule-transcending subject "whose skill has become so much a part of him that he need be no more aware of it than he is of his own body" applies to both of these (Dreyfus and Dreyfus, 1986, 30). This is what Dreyfus describes by showing how at the phenomenological level, the expert chess player and car driver are functionally equivalent *qua* expert. (Dreyfus and Dreyfus, 1986, 21-35)

Finally, Dreyfus's account provides a *practical expert point of view*. Based upon the universal scope of his model, he attempts to describe a common cognitive and affective relation to the world that all experts share, which can be evoked as the basis of justifying the expert's commitments. When other theorists have attempted to delineate the "expert point of view," they have mostly done so from a theoretically holistic perspective. For example, Scott Brewer argues that while nonexperts can know particular, true things about the methods and facts pertinent to a field, only experts know how the relevant features of a field, such as "enterprise" and "axiological" characteristics relate to provide a shared sense of what, how, and why practitioners in that field can claim to be true (1998, 1568-1593).[18] Brewer notes that his focus on the

---

[18] Brewer, like many other philosophers and legal theorists, appeals to the "point of view" as an analytical device that captures how perspective and justification relate (1998, 1568-1570). He turns to the point of view in order to articulate a common theoretical perspective that applies to all scientific experts—and that is arguably generalizable to all experts in all fields of expertise—regardless of which particular scientific field the practitioner is an expert in. He writes: "One invokes a point of view to justify some claim. To serve this justificatory function, the point of view is assumed to be a reliable method for achieving the (explicit or implicit) aims of some rational enterprise" (Brewer, 1998, 1575). Brewer's line of reasoning is that one turns to a point of view to rationally justify either a theoretical claim about what ought to be believed or a practical claim about what action ought to occur. What is distinctive about this type of validation is that it relates the justification of a claim to a distinct, yet "reliable" method, which is chosen to achieve a specific cognitive aim. The "reliable method" common to all "rational" points of view is defined in terms of two characteristics: "enterprise" and "axiological" conceptions. An "enterprise" is defined as the choice and use of particular methods of analysis in order to serve specific cognitive goals. He acknowledges that even within the "same generic enterprise" practitioners can disagree about the "proper specific aims of the enterprise" (Brewer, 1998,

organization of theoretical knowledge is somewhat Platonic (Dreyfus and Dreyfus, 1986, 1591). By contrast, Dreyfus often refers to
his skill model as anti-Platonic. Rather than providing a theoretically holistic account of the "expert point of view," he presents
a description of the common features of practical understanding.
The expert's practical understanding does not come from beliefs or
theoretical commitments, but from acquired and embodied skills
that provide the basis for determining whether rule following or
intuitive comportment are meaningful guides for acting in the field
one becomes expert in. Hence, Dreyfus writes, "The moral of the
five-stage model is there is more to intelligent behavior than calculative rationality". (Dreyfus and Dreyfus, 1986, 36)

4. Normative Implications

As detailed in the last section, Dreyfus's account is metaphysical
because he describes the essence of what experts can skillfully
do. This metaphysics suggests which expectations about the ser-
vices experts can and cannot render are justifiable, and what can

---

1571). But such a disagreement will take place within a "holistic" network involving an "axiological" component. When Brewer discusses the "axiological"
dimensions of justification, he does so in Larry Laudan's sense of the term.
When Laudan analyzes scientific reasoning, he distinguishes between "factual," "methodological," and "axiological" levels of analysis. Factual analysis
focuses on what exists in the world, including theoretical and unobservable en-
tities for scientists. At the methodological level, practitioners in a given field
share precise as well as vague rules. For scientists, this can include vague
rules, such as avoid ad hoc explanations, to precise rules, such as calibrate
instrument 'x' to standard 'y.' The axiological level, which is often explicable
in the form of rules, designates cognitive aims. Brewer, like Laudan, argues
that the relation between facts, methods, and axiological aims should not
be understood as a "simple linear hierarchy," but as "reticular" "constraints
[that] are multidirectional within the holistic network of aims, methods, and
beliefs" (Brewer, 1998, 1575). Facts, methods, and axiological aims relate in a
"multidirectional" sense because each can "constrain" the other, without any
one of the three being a priori valued more than the others. Facts, methods,
and axiological aims relate in a "holistic" sense since the point of view they
collectively constitute is comprised of all three features relating together. In
other words, a point of view is a complete and systematic perspective, irreducible to isolated observations. Finally, due to the holistic nature of a
point of view, Brewer describes its epistemic status as "understanding" and
not "knowledge." "The important difference between knowledge and understanding," Miles Burnyeat claims and Brewer repeats, "is that knowledge can
be piecemeal, can grasp isolated truths one by one, whereas understanding
always involves seeing connections and relations between the items known"
(Brewer, 1998, 1591).

and cannot be legitimately asked of them. Dreyfus's account thus has normative implications, which he discusses empirically in connection with, among others, ballistics examiners, chicken sexors, citizens, judges, nurses, paramedics, physicians, science advisors, and teachers (Dreyfus and Dreyfus, 1986, 196-201; Dreyfus, 1997, 106). A phenomenological understanding of the nature of expert decision making, Dreyfus suggests, needs to be the basis for identifying the means by which we may achieve some our social and political goals that involve experts, such as consulting experts for personal, institutional, and legal reasons. Although directly deducing normative obligations from a descriptive foundation might suggest that Dreyfus commits the naturalistic fallacy of deducing obligations about what ought to be done from premises that only state what is the case, he contends that the relation between phenomenology and normativity is an issue of "priority."[19]

These normative implications are significant for several reasons. A first reason is the renowned difficulty experts have communicating with others. Although this is sometimes attributed to the disparity between technical and ordinary language, and sometimes to psychological factors such as arrogance, Dreyfus's account suggests deeper causes, which I will return to shortly. Another reason the normative implications of his account are significant is the frequency with which experts serving as personal, institutional, and legal advisors are challenged as to their motives and biases, which may arise in connection with: (1) their professional training, which can influence how experts conceptualize and bound the problems they deal with; (2) their employers, which can influence politically the conclusions experts arrive at; (3) economic interests, which are capable of turning experts into hired guns whose recommendations can be purchased for the right price; and (4) the desire for recognition and reputation. By contrast, Dreyfus suggests that essential normative restrictions surrounding expert performance

---

[19] Dreyfus argues that the relation between phenomenology and normativity concerns "priority," in the sense that normative obligations are ascertained by phenomenologists "prioritizing" how agents do in fact respond to concrete situations. This "priority" is also, according to Dreyfus, held by those in the Hegelian tradition of *Sittlichkeit*, such as Bernard Williams, Charles Taylor, and Carol Gilligan. By contrast, Dreyfus argues, a group of theorists in the Kantian tradition of *Moralität*, such as Jürgen Habermas, John Rawls, and Lawrence Kohlberg, "prioritize" detached principles that detail what the right thing to do is over an understanding of the empirical conditions that allow for certain decisions to be made (1990, 237).

can be determined without considering the social forces that have sometimes influenced what claims experts make.

The normative implications arise from the claim that novices cannot transcend rules through developed intuition in the way that experts can in any field, no matter how dire or pressing the social circumstances might be.[20] It is therefore illegitimate, according to Dreyfus, to ask them to describe their process of decision making in propositional statements, because their decisions are made on the basis of tacitly operating intuition. Not only is the chess grandmaster acting on intuition, but so too is the ballistics expert who cannot propositionally express how he or she determined whether or not a particular bullet originated in a particular gun (Dreyfus and Dreyfus, 1986, 199). In assuming the role of expert legal witness the ballistics expert will be expected to correlate conclusions to rules, and only in doing so will he or she potentially be convincing to the jury. But Dreyfus argues this persuasive power comes at the expense of prioritizing the "form" over the "content" of the explanation (Dreyfus and Dreyfus, 1986, 199). In other words, the expert "forfeits" expertise when explaining his or her decision making process based on rules that are longer followed (Dreyfus and Dreyfus, 1986, 196). In appealing to rules the expert "forfeits" the intuitive experience that guided his or her decision making process.

Ultimately, Dreyfus argues that rational reconstruction of expert decision-making is an inaccurate representation of a process that in principle is unrepresentable. When nonexperts demand that experts walk them through their decision making process step by step so that they can follow the expert's chain of deductions and inferences (perhaps hoping to make this chain of deductions and inferences for themselves), they are, according to Dreyfus, no

---

[20] This is an important corrective, for instance to Paul Feyerabend, who for example, misses this point. He argues that when the prestige of science does not demand excessive complication, and social circumstances are such that experts cannot be overly esteemed, such as during wartime, medicine is capable of being effectively simplified. He claims that evidence for this can be found in army recruits who have historically proven themselves capable of being instructed as physicians with only half a year of training (Feyerabend, 1987, 307). The point that Feyerabend misses is that army recruits may quickly become instructed to be competent at various aspects of medicine, such as triage, but this training does not produce expert physicians or undermine how normal training allows experts physicians to do more than quickly trained ones.

longer allowing the expert to function as expert, but instead, are making the expert produce derivative representations of his or her expertise. Hence Dreyfus argues that too much pressure should not be placed upon experts to "rationalize" their "intuitive" process of decision making to nonexperts:

> It is often desirable that experts defend their recommendations against other experts, or in some way be cross-examined so that those effected can question their presuppositions. If this is taken to mean that the expert must articulate his values, rules, and factual assumptions, examining becomes a futile exercise in rationalization in which expertise is forfeited and time is wasted. (Dreyfus and Dreyfus, 1986, 196)

Dreyfus's normative position thus amounts to what Douglas Walton calls "a strong form of the inaccessibility thesis" (henceforth IT): "that expert conclusions cannot be tracked back to some set of premises and inference rules (known facts and rules) that yield the basis of expert judgment" (Walton, 1997, 110). According to IT, when experts render a verdict in matters of their expertise, their judgments are inaccessible to nonexperts in the sense that experts cannot propositionally express "a set of laws and initial conditions (principles and facts) that would exhibit an implication of the conclusion [i.e. what the expert concludes] by deductive (or even inductive) steps of logical inference" (Walton 1997, 109). Due to the expert's reliance on inexpressible dimensions of intuition, IT suggests that "the most advanced expert in a field, who has achieved outstanding mastery of the skills of her field, may be the least able to communicate her knowledge" (Walton 1997, 113). It is not merely that experts will be unable to communicate how their knowledge was achieved to nonexperts, but that their knowledge will be equally opaque to themselves and other experts.

At bottom, Dreyfus's chief normative concern is not that experts can mislead nonexperts, but rather that experts are "endangered" by the constraints they are sometimes subjected to: "Experts are an endangered species" (Dreyfus and Dreyfus, 1986, 206). He characterizes this as a problem confronting the U.S. in particular:"Demanding that its experts be able to explain how they do their job can seriously penalize a rational culture like ours, in competition with an intuitive culture like Japan's" (Dreyfus and Dreyfus, 1986, 196). Furthermore, Dreyfus argues that historical changes have precipitated the expert's current crisis. He contends that recent advances in computer technology and bureaucratic social organization exacerbate this cultural problem: "The desire to rationalize society would have remained but a dream were it not for the invention of the modern digital computer. The increasingly bureaucratic nature of society is heightening the danger that in the future skill and expertise will be lost through overreliance on rationality" (Dreyfus and Dreyfus, 1986, 195). In order to solve the problem of the disappearance of expertise Dreyfus seeks to reeducate people on the difference between "knowing how" and "knowing that." The future that Dreyfus speaks of is not distant. He suggests that "within one generation" we may lose "our professional experts" (Dreyfus and Dreyfus, 1986, 206). The goal is to get people to appreciate the value of intuition and the limits of rational deliberation:

> Society ... must encourage its children to cultivate their intuitive capacities in order that they might achieve expertise, not encourage them to become human logic machines. And once expertise has been attained, it must be recognized and valued for what it is. To confuse the common sense, wisdom, and mature judgment of the expert with today's artificial intelligence, or to value them less highly, would be genuine stupidity. (Dreyfus and Dreyfus,1986, 201)

5. Problems with Dreyfus's Descriptive Account

Dreyfus's account, however, admits certain categories of people as experts which do not belong, and omits several which do.Dreyfus's claim, for instance, that adults are 'experts' in walking and talking—which as we have seen is essential to his account for it grounds them in the same lifeworld spectrum as conventionally expert behavior—collides with ordinary usage. We do not call people who are merely ambulatory or verbal 'expert' walkers or talkers, but reserve the adjective for those who undergo special training, give professional advice, etc. We do not call licensed drivers 'experts'— nor even driving enthusiasts or competitive amateurs—even when they have an intuitive relation to operating their vehicles. Rather, we reserve the word for drivers who belong to professional driving organizations, participate in certain kinds of competitions, and so forth. In fact, there are many non-institutionally affiliated racing aficionados whose intuitive relation to skills may very well be developed in the manner Dreyfus describes. We call these people 'enthusiasts,' 'weekend warriors,' and sometimes even competitive 'amateurs,' but not experts. The point is that in principle an 'enthusiast' can be as skilled as an 'expert,' yet for social reasons we do not refer to him or her as an expert. What Dreyfus fails to recognize, is that his references to experts are descriptively loaded because he uses *socially acknowledged* experts as data for *asocial* phenomenological descriptions of expertise.

Meanwhile, Dreyfus's account also excludes certain classes of people from being experts who do belong. To appreciate this point, it is necessary to distinguish between 'expert x,' which is an adjectival use of 'expert' that stems from the Latin *expertus*, and 'expert in x,' which substantively treats 'expert' as a noun. In Dreyfus's terms, 'Expert x' corresponds to "knowing how" while 'expert in x' corresponds to "knowing that." Whereas an 'expert x' could be an 'expert farmer,' an 'expert in x' could be an expert "in farming." An expert "in farming" could effectively communicate, coordinate, and synthesize accurate propositional information about farming—could become Secretary of Agriculture—even if terrified of plows and tractors. An 'expert in sports,' who correlates the past behavior of athletes to current situations, could be crippled and lack physical capacity to play the sport; an 'expert in music' could be a terrible musician. Nevertheless, Dreyfus denies that the propositional knowledge definitive of an 'expert in x' is expert knowledge. He presents the following unflattering description of an 'expert in x' as a "kibitzer":

> Listening to... commentators, who take up at least half the time on erudite talk shows, is like listening to articulate chess kibitzers, who have an opinion on every move, and an array of principles to invoke, but who have not committed themselves to the stress and risks of tournament chess and so have no expertise. (Dreyfus, Spinosa, and Flores, 1997, 87)

Dreyfus associates the knowledge of 'experts in x' ("commentators" or "kibitzers") with the "idle talk" definitive of the "public sphere" that "undermines commitment stemming from practical rationality" (1997, 86). The reasoning of an 'expert in x' is "not grounded in local practices" but rather in "abstract solutions" and "anonymous principles" that fail to display "wisdom" (1997, 87). Moreover, Dreyfus argues, an 'expert in x' lacks the bodily commitment a genuine 'expert x' possesses; only an 'expert x' affectively cares about the outcome of a situation and experiences the "risk" of performance. Dreyfus's dismissal of an 'expert in x,' thus resembles a type of criticism sometimes leveled against art and music critics. According to this type of assessment, critics threaten the creative activity of artists and musicians because their derivative interpretations of creative activity can distort the meaning of that activity and socially influence which types of creative activity can flourish.

Dreyfus is wrong to deny an 'expert in x' the status of expert and should instead realize that an 'expert in x' simply is a different kind of expert than an 'expert x.' I believe he might be willing to change his position if he were willing to conceded that the commentary given by an 'expert in x' can strengthen rather than detract from a practice. Contrary to Dreyfus's claims, good expert commentators do more than "kibitz" and evoke abstract principles to criticize the primary participants of a practice. In many instances, an 'expert in x' does not detract from what an "expert x" does in a field, but rather an 'expert in x' helps spectators appreciate aspects of the practice that an 'expert x' engages in. For example, someone who is an 'expert in art history' does not necessarily detract from the paintings and sculptures made by an 'expert artist,' but often can aid art appreciators to learn what aspects of a painting and a sculpture are impressive. The same is true in the sciences where an 'expert in science,' such as a popularizer, can use ordinary language and striking images to present an overview that aids nonscientists to appreciate scientific developments. Furthermore, an 'expert in science' can address how science interacts with society. He or she can clarify what makes certain aspects of science controversial, noting how disputes between scientific experts develop, maintain themselves, and affect public policy.

The critical point is that an 'expert in x' is socially recognized as performing an established and valuable mediating role because his or her commentary is not, pace Dreyfus, always negative and judgmental. Without the critical commentary of an 'expert in x,' nonexperts might not appreciate why certain artistic and scientific activity is worth supporting. Don Ihde makes a similar point about the social value of an 'expert in x.' Additionally, he suggests that good criticism stems from an 'expert in x's' affective comportment:

> Few would call art or literary critics "antiart" or "antiliterature" in the working out, however critically, of their products. And while it may indeed be true that

given works of art or given texts are excoriated, demeaned, or severely dealt with, one does not usually
think of the critic as generically "antiart" or "antilit-
erature." Rather, it is precisely because the critic is
passionate about his or her subject matter that he or
she becomes a "critic". (Ihde, 1998, 127-128)

In contrast with Dreyfus, Ihde describes an 'expert in x' as "passionate" and not "detached." An 'expert in x' can be so passionate
about his or her subject matter that the process of writing a review can be fraught with all the emotions and risks that Dreyfus
believes only an 'expert x' experiences.

Another category of expertise passed over by Dreyfus's account
of skill acquisition is the "coach." The expert performer and the
coach or 'expert in x,' to be sure, are functionally different. Nevertheless, they are closely related enough to suggest that Dreyfus
is relying on a false dichotomy between those who "do" and those
who comment, kibitz, or at best instruct as a propadeutic to "doing." Coaching not only helps the learner get his or her body to
either approximate an ideal, standardized technique, but sometimes, aims at getting the learner to reach a more effective, yet
personalized bodily comportment. When successfully coached, the
learner's skills function in the most economic way possible for that
learner. Since these adjustments differ from body to body, they
cannot even be reduced to rules of thumb, which contain a higher
degree of generality than adjustments do. Although Dreyfus does
not acknowledge it, coaching is so important that many experts,
such as professional athletes, musicians, and actors, continually
seeks out advice from coaches in order to maintain their expert
performance. Even expert chess players seek out coaches not to
"kibitz," but to have their games evaluated and perhaps even
strengthened. By contrast, the instructor, who only dispenses
rules in Dreyfus's five-stage model of skill acquisition, altogether
disappears after the initial stages of learning. Even in Dreyfus's
revised model, one either learns by copying a "master," but not
by being coached, or by surveying different styles that different
"masters" possess, and then in some unspecified way coming to
develop one's own "style" (2001, 43-46). Thus, by overlooking
the role of coaching in the process of skill acquisition, Dreyfus es-
teems an 'expert x' as more *self-sufficient* than is warranted, and
demeans an 'expert in x' to an undeserved 'nonexpert' status.

The problem of 'self-sufficiency' extends not only to Dreyfus's
underdeveloped account of how people learn, but also to his mis-

understanding of how self-sufficient the body is when it adapts
to learning new skills. Like Merleau-Ponty's notion of the "lived
body," the body that goes through Dreyfus's stages of learning has
no influential biography, gender, race, or age (Young 1998; Sheets-
Johnston 2000). He acknowledges that "cultural styles" affect how
skills are learned, noting for example that differences exist between
how American and Japanese mothers "handle" their babies (2000,
46-47). However, this notion of "cultural style" is not very concrete since it depends upon unsubstantiated generalities and ignores crucial biographic differences that exist within individuals
sharing a single culture. Bobby Fischer's chess playing intuition,
for example, was not solely developed by playing many games of
American style chess. Some biographers contend that Fischer's
personal childhood experiences shaped the aggressiveness of his
style of chess playing. Likewise, a racist cop in an American pre-
cinct does not develop an intuitive sense of whether someone is
suspicious solely by having lots of committed on duty experience
observing suspicious criminals. Instead, a cop with racist tendencies intuits whether someone is behaving suspiciously on the basis
of familial or regional prejudices he or she absorbed about a racial
group, such as African-Americans, along with experiences in sus-
pect identification acquired from years on the force. The critical
point is that from Dreyfus's perspective, one develops the affective
comportment and intuitive capacity of an expert solely by immersion into a practice. He implicitly posits the body, which is the
locus of intuition, as immune to being shaped by forces external
to the practice one apprentices in. But this is a mistake, since
the affective demeanor and intuitive capacity that manifest in a
particular practice are influenced by a variety of factors, including biographic ones, even when one is committed to progressing
in a particular field. The apprentice's body is not self-sufficient
enough to develop its skills in isolation from broader social influences that first do not appear directly connected with the practice
one desires to become an expert in.

   What these comments on self-sufficiency suggest, is that Dreyfus's account lacks hermeneutical sensitivity. The flaw in his assumption that skilled behavior crystallizes out of contextual sensitivity plus experience without contribution from individual or
cultural biography can be traced to a failure to take into account
the fact that the embodied subject, *even when behaving expertly,*
brings to the situation what has been historically and culturally
transmitted to it, and in a way such that the subject can never

grasp cognitively all at once. The individual expert performer, as a consequence, does not have a complete purchase on his or her own expert behavior. Therefore, *contra* Dreyfus, it will always be possible in principle for an expert performer to learn about one's own performance from another, contextually sensitive person—though, again, this is not because the other has managed to obtain an objective position, but on the contrary, because the other is differently situated.

6. Problems with Dreyfus's Normative Account

Lack of hermeneutical sensitivity also affects Dreyfus's normative account, for his assumption of the autonomy of expert training suggests a naivete in his counsel to "trust experts." While a beginner might might have entered a training program *culturally* or *situationally embedded* with prejudices, ideologies, or hidden agendas, these would all be left behind by the time one reached the expert stage; an expert's knowledge for Dreyfus, as we saw, crystallizes out of contextual sensitivity plus experience. But if one can never leave the hermeneutic circle, the best that one can do is transform ones embeddedness, rather than extricate oneself from it. The acquisition of expertise is not a transcending of embeddedness and context, but a deepening and extension of one's relationship to it (Crease, 1997). This also means that experts will never be able to free themselves a priori from the suspicion that prejudices, ideologies, or hidden agendas might lurk in the pre-reflective relation that characterizes expertise.[21] Not only is

----

[21] At one point, Dreyfus suggests that trust can be obtained legitimately if experts provide a certain type of narrative. When he makes this suggestion, Dreyfus seems to mitigate his IT thesis. He suggests experts can be effective communicators, so long as they do not need to provide deductive accounts of rules they followed when making decisions: "The cross-examination of competing experts in an intuitive culture might take the form of a conflict of interpretation in which each expert is required to produce and defend a coherent narrative which leads naturally to the acceptance of his point of view" (Dreyfus and Dreyfus, 1986, 196). This passage is interesting because it suggests that experts are capable not merely of producing, but "defending" something called "coherent narrative" and that this narrative may be evaluated in an "intuitive culture" without endangering expertise. The passage is problematic because Dreyfus fails to explain what a "coherent narrative" is and why it is so efficacious as to "lead naturally" (another undefined phrase) to acceptance of a "point of view." The implication is that somehow experts can avoid the problems of IT, which are connected with what I called a "practical point of view," by presenting another kind of viewpoint in their "coher-

this suspicion to be expected; its absence would be socially dangerous. This gives rise to a *recognition problem* that is such a prominent part of many actual controversies involving a technical dimension where expert advice is required. But there is no room for this recognition problem in Dreyfus's account. He leaves no grounds for understanding how an expert might be legitimately challenged (or instructed, for that matter, as in the case of sensitivity training, nonexpert review panels, etc.). One would never imagine, from Dreyfus's account, that society could possibly be endangered by experts, only how society's expectations and actions could endanger experts. The stories of actual controversies, I will argue, not only shows things do not work the way Dreyfus says they do, but also that it would be less salutary if they did. Such stories, I claim, amount to a counterexample to Dreyfus's normative claims, and point to serious shortcomings in his arguments.

Dreyfus, let us recall, assumes from the start that the people who possess expert level skills are the same people who *should* be socially recognized as experts. It is unproblematic, for him, who "counts as" an expert in a given social situation; he assumes the absence of a recognition problem. This is clear from his examples. On the one hand, he refers to people who are socially recognized as experts, such as airplane pilots, surgeons, and chess masters, to illustrate how embodied expert performance functions (Dreyfus and Dreyfus, 1986, 30-35). These references are descriptively loaded because they use *socially acknowledged* experts as data for *asocial* phenomenological descriptions of expertise. On the other hand, he portrays mundane examples of everyday action, such as

---

ent narrative." The word "coherent" can be taken to suggest holism, and it is possible that Dreyfus has in mind something like Brewer's "theoretically holistic" expert point of view. But if Dreyfus is evoking theoretical holism, then he fails to explain how he can expect that experts, who according to IT "forfeit" their expertise when producing propositional content, are capable of providing such a narrative *qua* experts. The guiding presupposition is that whatever this "coherent narrative" is, it is only unproblematic for experts to produce in an "intuitive culture," such as Japan. Not only does this presupposition seem to be predicated upon unsupported cultural essentialism, but it also calls RT into question. Nonexperts may not be able to recognize experts on the basis of deductive procedural steps, but Dreyfus indicates recognition can occur on the basis of a "natural" response to the content of a "coherent narrative." Without explaining what these are, Dreyfus begs the question as to why nonexperts should trust expert intuition.

driving a car, walking, talking, and carrying on a conversation, as
paradigmatic instances of how experts behave, even though these
activities would not normally be socially recognized as being performed by experts (Dreyfus and Dreyfus, 1986, 30). By arguing
that the *same type* of expertise exists in both extraordinary performances of skill that are socially recognized as occurring at the
expert level and mundane performances of skill that are not recognized as occurring at the expert level, Dreyfus advances his end
of demystifying the seemingly magical quality of expertise and
establish continuity between expert and everyday lifeworld activity. In both instances, for him, expertise is a pre-reflexive relation
between skill and environment: "An expert's skill has become so
much a part of him that he need be no more aware of it than his
body". (Dreyfus and Dreyfus, 1986, 30)

From Dreyfus's perspective, social problems can arise when social agents are unable to recognize the essential qualities of expertise that are rendered explicit by the *phenomenological* invest-
igator. These social problems arise from a different source than
expertise itself; they are brought about by the *interference* of
political, social, and cultural involvements. But the hermeneutic
circle would turn this around; these involvements *give rise to* recognizing and trusting experts in the first place. Even if one defers
to the traditional "experts," that deference, too, has been brought
about as a result of one's particular background and lifeworld involvements. For expertise is a two- way relation: the *claim* to
expertise itself involves a social demand; it is not merely a neutral
identification label but a declaration that others should defer to
the expert's judgments. The phenomenon of expertise, therefore,
is ultimately and inextricably tied to its social utility; an expert
is not only "in" a field but "for" an audience.

An obvious example is the 'selection' problem of experts. In
many instances experts who endorse different conclusions within
the same field can be pitted against one another as 'counter experts.' A common strategy for 'counter experts' consists in claiming that the judgment proffered by the expert is tainted due to
the presence of prejudices, ideologies, or hidden agendas. 'Counter
experts' are particularly prevalent in the legal arena where expert
witnesses function in accord with the logic of an adversarial system
by proffering testimony for both the prosecution and defense.[22]

---

[22] As Shelia Jasanoff points out, the issue of selecting real experts is so
difficult in courtroom that the original 1923 "general acceptance" criteria

Stephen Turner's account of expertise addresses the recognition issue by pointing out that, in order for someone to be an expert, he or she needs not only to be skilled, but also to have an audience that socially recognizes his or her type of skills as skilled expertise (Turner, 2001, 138).[23] Turner contends that although what Merton calls "cognitive authority" is neither an "object that can be distributed" nor something that can be "simply granted," it nevertheless is "open to resistance and submission" based upon the evaluations of different audiences (Turner, 2001, 128).[24] By contrast, Dreyfus, who defines an expert solely on the basis of skill acquisition and use, methodologically excludes the audience from the description of experts and expertise, defining an expert as one with the right affective comportment and intuitive response to the

---

employed in *Frye v. United States* had to be refined because it "did not provide guidance as to how much agreement was enough, or among who" (Jasanoff, 1995, 62). General acceptance, which is an implicit presupposition of RT, did not legally work because it failed to: (1) clarify the degree of consensus required for establishing general acceptance, (2) set guidelines for how the contradictory results produced by variations in boundary work should be resolved, and (3) determine how to weigh results provided by frontier as opposed to established science (Jasanoff, 1995, 62). Since general acceptance proved to be a vague and ineffective standard for judging expert consensus in the legal arena, by extension it is problematic for Dreyfus to implicitly connect IT with RT. They are also routinely used in journalism, where sensationalism is generated by showing that a problem is so complicated that experts cannot agree on how to solve it. But for Dreyfus, this is not an issue. Dreyfus overestimates the overall trustworthiness of experts because he analyzes them as a general category and therein fails to recognize that expert consensus is field specific. Since the level of expert consensus is field specific it follows that the trustworthiness of expert intuition is not something that can be addressed in Dreyfus's general terms.

[23] This does not mean that an audience of comparable scale recognizes every field of expertise. For example, experts in physics are more widely recognized as possessing expert skills than are theologians, who are only recognized by a particular sect, which Turner calls a "restricted audience" (Turner, 2001, 131).

[24] For example, when discussing the relation between massage therapists and recognition, Turner notes that some people feel they benefit from massage therapy, whereas others do not find the promise of massage therapy to be fulfilled. The massage therapist is thus only considered to be (or more strongly put, only is) an expert for the former audience: "So massage therapists have... a created audience, a set of followers for whom they are expert because they have proven themselves to this audience by their actions" (Turner, 2001, 131).

situation.

This methodological move has many advantages over Dreyfus's phenomenology. It explains why: (1) expertise is not a stable property, but can be gained and lost; (2) why discussions of expertise tend not only to focus on epistemological, but also political issues; and (3) why perceptions of expertise can be based on historical transformations. In connecting expertise with these, Turner does not relinquish the philosophical aim of revealing general structures and transcending particularity (i.e. expertise can only be discussed with reference to local features). For example, he provides the following "taxonomy" that accounts for five general kinds of distinguishable experts:

> Experts who are members of groups whose expertise is generally acknowledged, such as physicists; experts whose person expertise is tested and accepted by individuals, such as the authors of self-help books; members of groups whose expertise is accepted only by particular groups, like theologians whose authority is only accepted by their section; experts whose audience is the public but who derive their support from subsidies from parties interested in the acceptance of their opinions as authoritative; and experts whose audience is bureaucrats with discretionary power, such as experts in public administration whose views are accepted as authoritative by public administrators. (2001, 140)

Turner's account thus relates the phenomenon of expertise to changing historical and social perceptions. Dreyfus might well retort that there is no reason why his phenomenological model should be understood as incompatible with Turner's—that Turner simply expands on Dreyfus's model, with Dreyfus describing what expert skills are and Turner describing how different audiences come to recognize these skills as occurring at the expert level. But the critical methodological difference is that Turner does not treat the possession of skill as a necessary *and* sufficient condition of expertise, omitting reference to an audience, while Dreyfus does.[25]

Dreyfus's failure to address the recognition problem is high-

---

[25] One may nevertheless be concerned that by insisting upon recognition as an essential dimension of expertise Turner undermines the objectivity of expertise. History is replete with examples of people who at one point are acknowledged to be experts and at other moments are denounced as charlatans.

lighted by the following paradox. On the one hand, he argues that only an expert can recognize another person as a genuine expert, for nonexperts do not know what to look for when evaluating whether someone is skilled.[26] It takes an expert—and only one—to know one. This might be called a *difference* claim (DC): experts are "not like us." DC is fairly innocuous, and even Turner adheres to a version: "[It] is the character of expertise that only other experts may be persuaded by argument of the truth of the claims of experts; the rest of us must accept them as true on different grounds than other experts do" (2001, 129). On the other hand, Dreyfus also advances a *similarity* claim (SC), according to which "they"—the experts—are very much "like us." They behave in a similar way the rest of us do in our everyday activities. At the most basic level of everyday coping, everybody deserves to be characterized as an expert: "Citizens will be speaking in terms of their expertise, whether they are university professors who have expertise in foreign cultures doing business with their state or farmers or small-store owners speaking about concrete problems that need legislative solution" (Dreyfus, Spinosa, and Flores, 1997, 107). Thus we can projectively identify with experts, and understand the kind of knowledge they use in their judgments. The problem is not only that Dreyfus needs to advance both, it is that, in the end, he needs SC to trump DC. Otherwise, nonexperts would lack any basis to recognize, accept, and trust the kind of knowledge that experts possess.

This point can be exposed most directly by reference to the

Turner acknowledges, "[What] counts as 'expert' is conventional, mutable, and shifting, and that people are persuaded of claims of expertise through mutable, shifting conventions" (2001, 145). Nevertheless, he claims that the prerogative to revise fallible judgments is a crucial part of democratic life and that to insist on a higher standard is "utopian" (Turner, 2001, 146).

[26] Again, it is helpful to consider Dreyfus's examples. An example of DC can be found in Dreyfus's description of an experiment in which students, experienced paramedics, and CPR instructors watched videotapes of six exemplars (five students and one experienced paramedic) giving CPR to patients. This target group was then asked which of the exemplars he or she would choose to save his own life. Dreyfus writes: "The results were revealing. In the paramedic group, nine out of ten selected an experienced paramedic. The students choose the paramedic five times out of ten. The instructors, attempting to find a paramedic by looking for the individual closely following the rules they were taught, failed to find the expert because an experienced paramedic has passed beyond the rule-following stage!" (Dreyfus and Dreyfus, 1986, 201).

situation depicted in Ibsen's play *Enemy of the People*. The central character, Thomas Stockmann, is a doctor at a spa on which the livelihood of his town depends. He thinks an invisible poison is polluting the spa's water, and confirms it by sending water samples to an expert at the University for analysis. He seeks to inform members of the community, thinking they will thank him for bringing the danger to their attention. But a friend warns him not to be so sure how they will respond: "You're a doctor and a man of science, and to you this business of the water is something to be considered in isolation. I think you don't perhaps realize how it's tied up with a lot of other things." Stockmann, citing the expert, insists that the "shrewd and intelligent" people will be "forced" to accept the news. The mayor points out that, for citizens, the matter is more economic than scientific; and indeed, the citizens condemn Stockmann as an "enemy of the people." The basic conflict thus involves a scientist who accepts expert technical advice as authoritative, versus citizens who do not find that expert advice authoritative, who find it threatening to their world, and who seek guidance from others.

Ibsen's invented situation is like a model that strips away inessential details to clarify the essential forces of a situation. The situation involves a volatile controversy with a scientific-technical dimension in which people have lost confidence in traditional "experts." From the audience position, we see that two kinds of stories can be told about such a situation from two different perspectives. In the *expert's* perspective, that of Stockmann and the University chemist, there is no uncertainty or grounds for contestation regarding whom the expert is, what kind of training that person requires, and what kind of information (the technical issue of the quality of the water) is relevant to maximizing the good of that particular social situation. The experts are like the citizens in that they want to maximize the good of the town; they do so with their special knowledge. Failure of the citizens to recognize the expert, and defer to the expert's advice, is due to the peripheral, and even deleterious and corrupting, influence of economic and political motives and the involvements of the citizens with politicians, the media, and other nonexpert authorities. In the *citizen's* perspective, on the other hand, things are much more confused. The background and lifeworld involvements of citizens mean that economic and political motives loom much larger than they do for Stockmann and the chemist, and suggest different people to whose advice they should defer in seeking to advance their welfare.

In this perspective, the purported "experts"—Stockmann and the chemist—are precisely *not* like the other citizens for they have different agendas (and the chemist is even literally an outsider).

We see the difference between these two situated perspectives clearly from a *third* perspective, sitting in the audience. This third perspective is also not neutral, and its effect is to dramatize the similarity claim; we recognize the common humanity between Stockmann and the citizens, and the relevance of his knowledge to their welfare. One realizes that the welfare of the citizens requires that they defer to the experts, even as one fully appreciates the severity of the conflict and the impossibility of its resolution. But we in the audience have no doubt about who the real experts are, and the difference between essential and peripheral involvements. The person in the audience is not standing anywhere, not situated with respect to this aspect; this third position, in short, is not a hermeneutically sensitive one.

In any real controversy, however, no one occupies the audience position; everyone is as it were "on stage" in a situated position, standing someplace with *particular* involvements which gives rise to a *particular* understanding of the situation and, with it, an inclination to accept some people rather than others as authoritative. It is not a priori clear which involvements are essential and which particular; whose actions are in the grip of prejudices, ideologies, and hidden agendas, and whose are in the best interests of society. Meanwhile, experts are also in a particular position, standing someplace and with particular involvements, and the claim to expertise is a charge, a valence, a demand that one should be deferred to. Each real life controversy involving expertise takes the form of a jockeying between those who advance claims of expertise to advance their authority and those seeking the right authorities to whom to defer. This is the hermeneutic predicament: there is no escape from the particular involvements of a given situation. There is and can be no talisman for expertise.

This problem is most visible in connection with volatile public controversies involving a technical dimension—and especially involving public safety—where traditional sources of authority have become distrusted. In these cases, the question of who speaks authoritatively, of who is an expert, is contested. Each citizen, and each person proposed as an expert, has a particular set of involvements, and there is no safe audience position from which to sort out the essential and inessential involvements in an expert's judgment.

Consider, for instance, the case of the closing of the National Tritium Labeling Facility (NTLF) – where, as it happened, Dreyfus's wife Genevieve worked. The facility created unique tagged molecules by putting tritium into specific molecular positions, creating tracers that are used to study mechanisms of biochemical transformation in basic and applied research. But anti-nuclear activists objected to the fact that some tritium was released to the environment in the process. Scientists and local, state, and federal public health experts, after carefully examining the situation, said the emissions were safe, a fraction of the Environmental Protection Agency suggested limit, which in turn was a fraction of the background level. But the anti-nuclear activists sought to discredit those claims, saying that they were either made by those who worked at the facility, or from institutions connected in some way with that facility, or by scientists who knew too much about tritium to be disinterested. The activists were effective at disrupting the normal socially negotiated procedures for who speaks authoritatively about safety, and the facility was closed (Crease, 2002a).

Or again, consider the controversy surrounding the shipment of spent fuel rods from Brookhaven National Laboratory's High Flux Beam Reactor in 1976 (Crease, 1999). The controversy pitted activists, who associated research reactors with power reactors/nuclear weapons/the military, versus scientists for whom such associations made no sense at all. The situation spawned a spectrum of experts of the sort described by Turner's taxonomy. In yet another controversy involving the Brookhaven lab, a program it ran studying the health of Marshall Islanders accidentally exposed to fallout from a nuclear weapons test came under attack, with one complicating factor being that it involved the classic colonialist situation of U.S. scientists working in a third-world country whose language and customs they were not familiar with (Crease, 2003).

Each of these controversies was complex, and involves high stakes. Each turned on a scientific-technical issue, and thus necessary recourse to expertise. Yet who the "real" experts were deemed to be depended on who one was and where one stood. To understand such controversies involving experts and expertise requires moving beyond the practical expert's point of view.

7. Conclusion

Dreyfus's model of expert skill acquisition is philosophically important be- cause it shifts the focus on expertise away from its social and technical externalization in STS, and its relegation to the historical and psychological context of discovery in the classical philosophy of science, to universal structures of embodied cognition and affect. In doing so he explains why experts are not best described as ideologues and why their authority is not exclusively based on social networking. Moreover, by phenomenologically analyzing expertise from a first person perspective, he reveals the limitations of, and sometimes superficial treatment that comes from, investigating expertise from a third person perspective. Thus, he shows that expertise is a prime example of a subject that is essential to science but can only be fully elaborated with the aid of phenomenological tools.

However, both Dreyfus's descriptive model and his normative claims are flawed due to the lack of hermeneutical sensitivity. He assumes, that is, that an expert's knowledge has crystallized out of contextual sensitivity plus experience, and that an expert has shed, during the training process, whatever prejudices, ideologies, or hidden agendas that person might have begun with. This assumption not only flaws in Dreyfus's descriptive account but his normative account as well.

The phenomenological goal is to expose presuppositions that lurk unapprehended in the natural attitude. The phenomenological experience reveals that the most difficult presuppositions to expose and get a grip on are those closest to home. In this spirit, though possibly with a trace of perversity, one might expose this limitation by posing the following question: "Why does Genevieve Dreyfus no longer work at the National Tritium Labeling Facility?"

From Dreyfus's own account, it is because of a breakdown, a corruption of a legitimate and phenomenologically justifiable authority. A group of anti-nuclear crusaders managed to hijack a socially negotiated process and persuade the administrative authorities to overlook robust science and expert advice, and make a purely political decision. But from another, more all-too-common, highly important, and potentially powerful point of view, one could, with hermeneutic sensitivity, tell quite a different story. Individuals do not take in an item of information, even scientific information, nakedly. It matters who conveys that piece of information and in what context. The Berkeley anti-nuclear activists live in a different interpretive world than the scientists, with a different set of supporting behaviors, values, and institutions; their meaning-generating process by which they interpret facts, principles, and their application is very different (Crease, 1999, p. 498). The difference between the scientists and activists thus cannot be regarded as one between knowers and those who corrupt or betray that knowledge, but more like the relation between the members of one culture and another. Surely a description of this hermeneutic predicament, missing from Dreyfus's own, belongs in any account of expertise.

Finally, we believe that when evaluated from an immanent perspective, Dreyfus's account of expertise fails according to his own philosophical standards. In many publications Dreyfus champions Heidegger's hermeneutic approach to phenomenology over Husserl's. He even argues that Heidegger's account of authentic *Dasein* in Division II of *Being and Time* entails that *Dasein* is an "expert" (Dreyfus, 2000). However, by approaching trust as a static scenario in which a nonexpert solicits advice from an expert, Dreyfus resorts to an implicit Husserlian schema of intersubjectivity. The essence of SC is an analogy: just as I do not expect myself to be able to articulate rules for how to drive or ride a bicycle, I should not expect experts, such as ballistics examiners, nurses, and ecologists, to justify their decisions by referring rules. Dreyfus implicitly argues that even though trust is established at the social level on the basis of an expert's track record, at the phenomenological level, it is established through what Husserl calls "intersubjective pairing." By making a pre-reflexive analogy from my behavior to the behavior of an expert, I recognize that an expert's behavior is essentially similar to my own. The expert is an expert "alter ego." In the Dreyfusian scheme, I should trust experts because: (1) I trust myself to make decisions in a similarly intuitive manner, and (2) I trust my decisions to be good ones, even though I, like an expert, cannot propositionally justify them according to rules. Even though an expert has more training than I do, our cognitive similarities outweigh our technical differences.

Dreyfus's account of intersubjectivity here, with its lack of hermeneutic sensitivity, recalls Husserl more than Heidegger. Indeed, his portrait of the expert, who masters his or her own relation to expertise by making it through all of the developmental stages, and who feels no need to seek external ratification of his or her own abilities, evokes the caricature of Heidegger according to which authentic *Dasein* is accorded too much heroic freedom. But Heideggerian sensitivity to the hermeneutic dimensions of worldhood would have to depict an expert as engaged in a much more fragile and vulnerable process, in which expertise does not appear as a destination in which the individual surpasses ones embeddedness in that world. Expertise is always a process of becoming, and, in principle at least, it will always be possible for coaches, commentators, and others whose own expertise overlaps with the expert's to disclose aspects of expert performance which escape the grasp even of that performer.

Bibliography

Barbour, I. (1993). *Ethics in an Age of Technology*. San Francisco: Harper Collins.

Black, B., Ayala, F., and Saffran-Brinks, C. (1994). "Science and the Law in the Wake of Daubert." *Texas Law Review* 72: 715-802.

Brewer, S. (1998). Scientific Expert Testimony and Intellectual Due Process. *The Yale Law Review* 107 (4): 1535-1681.

Callon, M. (2001). Researchers in the Wild and the Rise of Technical Democracy. Paper presented at *Knowledge in Plural Contexts*, Science and Technology Studies, Université de Lausanne, Switzerland.

Caudill, D. and Redding, R. (2000). Junk Philosophy of Science? The Paradox of Expertise and Interdisciplinarity in Federal Courts. *Washington and Lee Law Review* 57 (3): 685-766.

Collins, H. M. (1995). Humans, Machines, and the Structure of Knowledge. *Stanford Humanities Review* 4 (2): 67-83.

Collins, H. (2001). Tacit Knowledge, Trust, and the Q of Sapphire. *Social Studies of Science* 31 (71-86).

Crease, R.P. (1997). Hermeneutics and the Natural Sciences: Introduction. In R. Crease (Ed.), *Hermeneutics and the Natural Sciences*, pp. 1-12. Dordrecht: Kluwer.

Crease, R. P. (1999). Conflicting Interpretations of Risk: the Case of Brookhaven's Spent Fuel Rods. *Technology*, 6: 495-500.

Crease, R. P. (2001). Anxious History: The High Flux Beam Reactor and Brookhaven National Laboratory. *Historical Studies in the Physical and Biological Sciences* 32 (1): 41-56.

Crease, R. P. (2002a). Compromising Peer Review. *Physics World*, 17.

Crease, R. P. (2002b). The Pleasure of Popular Dance. *Journal of the Philosophy of Sport*, 29 (2):106-120.

Crease, R. P. (2003). Fallout: Issues in the Study, Treatment, and Reparations of Exposed Marshall Islanders. In R. Figueroa and S. Harding (Eds),. *Exploring Diversity in the Philosophy of Science and Technology*, Routledge, 106-125.

Dreyfus, H. (1967). Alchemy and Artificial Intelligence. *Rand Paper P-3244*.

Dreyfus, H. and Dreyfus, S. (1985). From Socrates to Expert Systems: The Limits of Calculative Rationality. In C. Mitcham and A. Huning (Eds.), *Philosophy and Technology II: Information Technology and Computers in Theory and Practice*, pp.111-130. Boston: D. Reidel Publishing Company.

Dreyfus, H. and Dreyfus, S. (1986). *Mind Over Machine: The Power of Human Intuition and Expertise in the Era of the Computer*. New York: Free Press.

Dreyfus, H. and Dreyfus, S. (1990). What is Morality? A Phenomenological Account Of the Development of Ethical Expertise. In D. Rasmussen (Ed.), *Universalism Vs. Communtarianism: Contemporary Debates in Ethics*, pp. 237-264. Cambridge: MIT Press. Dreyfus, H. (1991). *Being-in-the-World: A Commentary on Heidegger's Being and Time*. Cambridge: MIT Press.

Dreyfus, H. (1992). *What Computers Still Can't Do: A Critique of Artificial Reason*. Cambridge: MIT Press.

Dreyfus, H., Spinosa, C., and Flores, F. (1997). *Disclosing Worlds: Entrepreneurship, Democratic Action, and the Cultivation of Solidarity*. Cambridge, MIT Press.

Dreyfus, H. (1998). Intelligence Without Representation. *Network for Non-Scholastic Working Paper*, Department of Philosophy, Aarhus University, Denmark.

Dreyfus, H. (1999a). How Neuroscience Supports Merleau-Ponty's Account of Learning. Paper presented at the *Network for Non-Scholastic Learning* Conference, Sonderborg, Denmark.

Dreyfus, H. (1999b). The Primacy of Phenomenology over Logical Analysis. *Philosophical Topics* 27 (2): 3-24.

Dreyfus, H. (2000). Could Anything be More Intelligible than Everyday Intelligibility? Reinterpreting Division I of *Being and Time* in the Light of Division II. In J. Faulconer and M. Wrathall (Eds.), *Appropriating Heidegger*, pp. 155-170. Cambridge: Cambridge University Press.

Dreyfus, H. (2001). *On the Internet*. New York: Routledge.

Feyerabend, P. (1987). *Science in a Free Society*. London: Verso.

Huber, P. (1991). *Galileo's Revenge: Junk Science in the Courtroom*. New York: Basic Books.

Huber, P. and Foster, K. (1999). *Judging Science: Scientific Knowledge and the Federal Courts*. Cambridge: MIT Press.

Husserl, E. (1973) *Cartesian Meditations and the Paris Lectures*, ed. S. Strasser. Dordrecht Kluwer.

Ibsen, H. (1988). *Ibsen: The Complete Major Prose Plays*, trans.

R. Fjelde. New York: New American Library Trade.

Ihde, D. (1998). *Expanding Hermeneutics: Visualism in Science.* Evanston: Northwestern University Press.

Jasanoff, S.(1995). *Science at the Bar: Law, Science, and Technology in America.* Cambridge: Harvard University Press.

Latour, B. (1999). *Pandora's Hope: Essays on the Reality of Science Studies.* Cambridge: Harvard University Press.

MacKenzie, D. (1996). *Knowing Machines: Essays on Technological Change.* Cambridge: MIT Press.

Mialet, H. (1999). Do Angels Have Bodies? Two Stories about Subjectivity in Science: The Cases of William X and Mister H. *Social Studies of Science* 29 (4): 551-582.

Pappas, G. (1994). Experts. *Acta Analytica* 9 (12): 7-17.

Polanyi, M. (1974). *Personal Knowledge: Towards a Post-Critical Philosophy.*Chicago:University of Chicago Press.

Rawls, J. (1968). Two Concepts of Rules. In J. Thomson and G. Dworkin (Eds.), *Ethics*, pp. 104-135. New York: Harper & Row.

Reber, A. (1995). *Implicit Learning and Tacit Knowledge: An Essay on the Cognitive Unconscious.* New York: Oxford University Press.

Rouse, J. (1987). *Knowledge and Power: Toward a Political Philosophy of Science.* New York:Cornell University Press.

Ryle, G. (1984). *The Concept of Mind.* Chicago: University of Chicago Press.

Schon, D. (1983). *The Reflective Practitioner.* New York: Basic Books.

Searle, J. (1983). *Intentionality: An Essay in the Philosophy of Mind.* New York: Cambridge University Press.

Searle, J. (1992). *The Rediscovery of the Mind.* Cambridge: MIT Press.

Sheets-Johnstone, M. (1999). *The Primacy of Movement.* Philadelphia: John Benjamins.

Sheets-Johnstone, M. (2000). Kinetic Tactile-Kinesthetic Bodies: Ontogenetical Foundations of Apprenticeship Learning. *Human Studies* 23: 343-370.

Stengers, I. (2000). *The Invention of Modern Science*, trans. D. Smith. Minneapolis: University of Minnesota Press.

Turner, S. (2001) What is the Problem With Experts? *Social Studies of Science* 31: 123-149.

Young, I. (1998). Throwing like a Girl. In D. Welton (Ed.) *Body and Flesh: A Philosophical Reader*, pp.259-273. Massachusetts: Blackwell Publishers.

Walton, D. (1997). *Appeal to Expert Opinion: Arguments from Authority.* University Park: Pennsylvania State University Press.

Williams, R. (1976). *Keywords: A Vocabulary of Culture and Society.* New York: Oxford University Press.

Winner, L. (1995). Citizen Virtues in a Technological Order. In A. Feenberg and A. Hannay (Eds.), *Technology and the Politics of Knowledge*, pp. 65-84. Bloomington: Indiana University Press.

Winograd, T. (1995). Heidegger and the Design of Computer Systems. In A. Feenberg and A.Hannay (Eds.), *Technology and the Politics of Knowledge*, pp. 108-127. Bloomington: Indiana University Press.

2

Chess-playing Computers and Embodied Grandmasters: In What Way Does the Difference Matter?

Introduction

As a theorist, I find chess interesting even though I do not play the game more frequently than once or twice a year. Chess narratives are seductive because they convey four recurring motifs concerning how people fundamentally use artifacts—

1. *Humans vs. Machines*: The fascination with computers competing against humans at chess attests to long-standing fears about artificial intelligence being superior to human intelligence, and algorithmic processing replacing intuitive judgment. As concern builds over the possibility that a "posthuman" future might arise, both religious and secular discussions express a palpable yearning for evidence that human excellence exists and can be recognized during the moments in which the so-called "human spirit" triumphs over its technological replacements.

2. *Humans vs. Other Humans*: The fascination with chess competition attests to perennial interest in how and why humans try to establish self-worth by besting other humans in games that require equipment to play.[1] Whether or not chess is an actual "sport" is irrelevant as far as this matter

---

[1] Of course, humans don't always try to establish self-worth in this way. Nevertheless, the use of equipment for this purpose has been a significant recurring historical motif.

goes. The crucial point is that chess is an adversarial endeavor, and it cannot be engaged without using the pieces and a board.

3. *Humans vs. Themselves*: The fascination with chess competition attests to perennial interest in how and why humans try to transform themselves by cultivating discipline in practices that require equipment to participate in. Chess is a potent example of this phenomenon because game play does not revolve around the contingent conditions that influence how other competitions are conducted (e.g., unfair referees, weather tarnished fields, unlucky dice or cards, *et cetera*). In this sense, players assume direct responsibility for their conduct, including their improvement, decline, and stagnation.

4. *Humans vs. Nature:* The fascination with chess competition attests to perennial interest in the question of how much transcendence the human mind is capable of. Not only do novices play the game in a wholly different manner than Grandmasters (i.e., the highest ranking international chess title), but chess is a domain that offers a symbolic glimpse into indefiniteness—new tensions and ideas always lurk on the horizon. No matter how well one plays, there is always more to learn.

The purpose of this chapter is to contrast two different *existential* portrayals of chess. One is provided by Hubert Dreyfus, a highly regarded philosopher of technology. The other comes from renowned director and screenwriter Ingmar Bergman, as expressed in his cinematic masterpiece *The Seventh Seal* (1957). Although Dreyfus focuses on motifs 1-4, and Bergman's exploration privileges the 4th issue, juxtaposing both treatments can shed light on the question: *What would it mean for computers to play chess like humans do?*[2]

---

[2] My colleague, Jesús Aguilar, objects to my analysis on the grounds that it focuses too intently on the contingencies of chess playing. In doing so, Aguilar contends, I fail to address chess itself. That latter, Aguilar insists, is a formal domain that can be instantiated in many ways—some of which might include far less embodied interaction than the ones I'm concentrating upon. In other words, to talk about the "game of chess" is to discuss all the possible conditions under which chess can be formalized and played. On this matter, I plead for phenomenological tolerance. I'm confining my interests to ways in

My central claim will be that, as a matter of contingency, human interaction is, *de facto*, interaction between agents in *culturally marked bodies*. Not only is Dreyfus insensitive to this point, but so too are chess commentators who fail to emphasize that *computers do not currently play chess like humans do because their strategy and tactics are not influenced by perceptions of culturally embodied persons who signify as perceived bodies—in terms of age, race, and gender, for example—and who signify performatively as they play—concentrate, deliberate, and make moves, for example*. What I will show, therefore, is that under the guise of rigorous phenomenological description, Dreyfus becomes blinded by a normative agenda that prejudices his account of what it is like for Grandmasters to play chess. By trying to establish why it is dangerous to replace human judgment with computerized processes, Dreyfus surreptitiously projects a *fantasy* in which *eros* and *prejudice*—two hallmarks of human disposition—fail to be capable of influencing how the Grandmaster perceives a move. In this respect, despite rhetorically touting the importance of "embodiment," Dreyfus unwittingly infuses his account with *disembodied images* of chess players who more closely resemble affectless computers than flesh-and-blood agents.

Before discussing Dreyfus, however, our attention will turn to Bergman. The cinematic analysis that follows will be largely *symbolic*, and as such, is presented in order *to orient the reader to some of the basic issues concerning embodiment*. The main point, which will be clearer as the essay progresses, is that Bergman highlights the problem of *bodily limitations* more explicitly than Dreyfus does. Indeed, the case will be made that if Dreyfus attended to embodiment as closely as and in the ways that Bergman does, then he would be better positioned to present a more accurate account of the type of perception that human chess players employ.

1. Bergman: Chess against Death

When Dreyfus analyzes how Grandmasters determine where to move their chess pieces, he relies upon insights from a philosophical tradition called existential phenomenology.[3] What existential

---

which chess has actually been played by humans who happen to play chess in this world with the bodies they happen to have. Thought experiments that conjure additional variations are, therefore, not being entertained.

[3] Existential phenomenologists focus on clarifying core dimensions of hu-

phenomenologists attempt to provide is an account of the invariant structures of experience—structures of experience that apply to all (or, at least, to all "typical") human beings by virtue of the fact that a shared human condition underlies all cultural differences. Bergman's *The Seventh Seal* can be interpreted as a cinematic expression of just such a philosophical ambition.

While Bergman uses many images to explore philosophical themes in the film, the now iconic motif of chess is, perhaps, privileged above all others. It serves as the central window through which the viewer can engage with Bergman's metaphysics—a reality in which coming to grips with God's absence can prove tormenting, and in which playing the game of chess is not a distraction from life, but rather a symbolic confrontation with the meaning of *finitude*.

Set in the Middle Ages when the ravages of the plague made it commonplace to think about mortality, the film opens with Antonius Block, a knight returning from the Crusades, challenging Death's personification to a chess match. Because Death concedes that, as the "paintings" and "legends" attest, he has a predilection for chess and is a rather skilled player, Block's request is readily accepted. Block may live for the duration of their game, and he may go free, should he prove capable of winning.

Rather than completing their contest in a single temporal sequence, Bergman's elaborate choreography allows for the allegorical quality of chess to linger throughout the film's duration. As each move unfolds, and as Block's early success turns into inevitable defeat, the viewer is provoked to engage in consideration of mortality. In this context, two scenes deserve to be highlighted. First, Block informs a priest that he plans on besting Death only to learn that the priest is, in fact, Death himself in disguise. Second, in a sequence that inaugurates the film's closure, Block distracts Death by knocking the pieces of the chessboard

---

man experience—not only the conditions under which we perceive, act, feel, and think, but also the sensual quality of such experiences, e.g., what it is like for human beings to actually undergo experiences of perception, action, feeling, and thinking. While many disciplines ranging from psychology to sociology and neurobiology also explore these features of mentality and embodiment, existential phenomenologists typically contend that alternative perspectives distort essential dimensions of "lived experience": some accounts allow theoretical conjecture of how experience should be organized to trump careful description of experience itself; others wish to reduce the significance of experience to the natural causes that produce such an experience.

over in order for a family of actors to escape undetected. We will return to both of these scenes shortly.

Considering the depth of Bergman's themes, one might wonder why he emphasizes chess.[4] As Woody Allen's short story "Death Knocks" cleverly shows, far from always suggesting profundity, it can be comical to imagine game playing as the defining test that determines whether one's life can be bargained for. In homage to Bergman, Allen describes a man trying to extend his life by playing Gin Rummy with Death. And yet, without being anachronistic, this parody (and the subsequent ones) can be viewed as evidence that it is unlikely that Bergmann expected the cliché of "life as a game" to suffice as a resonant expression of existential insights. Parodies tend to create comic effect only when they subject substantive ideas to imitation mixed with revision.

The power of Bergman's choice, then, likely depends upon viewers perceiving chess to be a unique activity—an archetypical bridge between Eastern and Western histories that occupies a special niche in the collective imagination. Like Bergman, numerous artists and thinkers have appropriated the motif of chess to explore fundamental dualisms: life and death, success and failure, active and passive comportment, defensive and offensive posturing, day and night, sacrifice and repayment, *et cetera*. For present purposes, however, the most relevant consideration that Bergman seems to evoke is the view that to play chess skillfully is to call forth intellectual powers that *define our humanity—both our singularity and our commonality with other natural beings*. To clarify these points, let us begin with the former consideration.

Chess playing requires foresight, and this, in turn, makes it incumbent upon players of varied levels of skill to consider *all three tenses of time*: not only present moves, but past and future ones as well. Extrapolations from this simple observation can remind us of criteria that have been proffered in order to demarcate humans from other beings, both organic and inorganic ones.[5]

---

[4] Clearly, the historical connection between chess and the medieval knight's seven virtues is a relevant background consideration. Nevertheless, the case would be overstated were one to claim that Bergman is engaged primarily with the presence of chess in knightly literature and practice.

[5] While these specific points of demarcation are not explicitly addressed in the manifest content of Bergman's film, the creative license that I'm taking here is justifiable on aesthetic grounds. Despite memorable dialogue and an unyielding thematic focus, the film's logic is not conveyed propositionally;

The first demarcation at issue is one that separates humans from things. Like human chess players, inanimate objects, ranging from chess pieces to chess playing computers, have temporal existence; but unlike humans, objects do not experience temporality as a theme for reflective contemplation. Indeed, inanimate objects have no sense of how they were created.[6] Neither do they have beliefs or desires about how to best spend the time that will elapse before they malfunction and ultimately disintegrate. For reasons such as these, the German phenomenologist Martin Heidegger calls things "world poor."

The second demarcation at issue is one that separates humans from animals. Like chess players, other animals have temporal existence; but unlike them, animals cannot experience the past and future as objects of critical reflection. To be sure, many animals have some sense of the past. In addition to evolutionary proclivities, direct experience with previously encountered predators, food sources, communal norms, and, in some cases, tools, influence how they currently perceive other creatures and beings. Moreover, that some animals will go to great lengths to preserve their own lives and the lives of their kin, attests to their ability to experience some connection to the future by virtue of how they express their primal aversion to threat.[7] However, according to many authorities, animals do not treat being born or dying as subjects that require explanation; given the constraints upon their language and cognition, it is doubtful that they can even entertain or express such thoughts. It is my understanding that despite the ability of some primates to learn to communicate through sign language, they only learn to signify *current* desires (e.g., hungry for a ba-

---

the issues it raises about human hopes and frustrations are disclosed through poetic *combinations of image and sound*—combinations that elicit existential thoughts from many viewers precisely because some of the ideas they call to mind are not themselves displayed didactically, as, for example, clear and distinct lessons put forth for viewers to passively absorb.

[6] My discussion of the limits of "things" is meant to convey the essence of Heidegger's position. More contemporary theorists who study digital technologies, cybernetic systems, and "quasi-objects" could, of course, complicate this account of what things can and cannot do.

[7] My colleague David Suits claims that only humans can have an aversion to death. Non-humans might, he suggests, primarily experience a primal aversion to pain in the scenario detailed above. Due to the focus of the essay, this question will need to be explored in detail elsewhere.

nana now)—past desires (e.g., was hungry for a banana yesterday) and future desires (e.g., will be hungry for a banana tomorrow) are not part of their communicative repertoire. For reasons such as these, Heidegger claims that an "abyss of essence" separates humans from the other animals, making them, thus, along with things, "world poor." And whereas their primal relation to death allows for animals to "perish," it is the human capacity to organize existence around an understanding of lifespan that, according to Heidegger, allows us, and only us, to experience "death" at the end of our days.

Having just detailed a chain of associations running from chess to temporality to the unique human relation to death, we can now turn to the second issue that Bergman's treatment of chess evokes—namely, the theme of humans as natural beings. This theme is raised at the margins of the film's examination of cognitive limitations. Here's how

In addition to temporal considerations, the game of chess also provides an occasion for players of varied skill to reflect upon their *cognitive limits*. This is because learning how to play chess—as well as learning how to improve upon past chess performance—is a process that requires players to develop a sophisticated grasp of how to *think about thinking*. As FIDE Master Amatzia Avni notes, by raising introspective questions, such as "How should I tackle this problem?" and "What strategy of problem-solving should I adopt?", players increase their *meta-knowledge*. They learn, in other words, how to estimate what they know and how to discern the extent to which their awareness is limited (Avni 2002, 62-63).

Because such inquiry into the *relation between knowledge and ignorance* can allow for chess players to appreciate how their subjectivity is biased, the process evokes the ancient Greek injunction, "Know thyself!" Although Socrates is said to have uttered these words in order to convey his conviction that the unexamined human life is not worth living, a parallel to chess is still warranted. The unexamined chess game is not worth playing because one of the game's deep allures is its potential to reveal a player's finitude. Put otherwise, inquiry into meta-knowledge can provide a beneficial antidote to *hubris* because it has the capacity to remind participants that as finite and embodied beings, all chess players are: subject to distraction and fatigue, capable of misinterpreting overt and subtle forms of behavior, and prone to under- and over-preparing for foreseeable conflict. It is the arrogant or naive person, the one who believes that embodiment can be transcended, that experiences the resistances offered by life and chess as painful and frustrating instead of as essential limit conditions that make struggle a fundamentally meaningful human activity. For reasons such as these, Bergman's embodied depiction of Death is a masterfully poetic image; it allows him simultaneously to portray Death as a supernatural force and as an embodied presence who, *qua* embodied being, is distractible.

What, then, could be a more appropriate cinematic vehicle for Bergman to have selected in order to raise questions about the limits of the human intellect—about whether, in the end, all matters of scheming and planning are expressed through our embodied limitations? In short, Bergman might have focused upon chess playing to convey existential insights because the game can be understood as a *memento mori*—a reminder of fragility and mortality. As with even the best games of chess, the most rousing lifestyles remain imperfect and cannot endure forever. Even if humans experience death differently than animals do, both remain organic beings—and as such, both experience life through the matrix of bodily imperfection and, ultimately, its cessation through bodily dissolution. To return to the first of the memorable scenes highlighted earlier, even priests cannot forestall the inevitably of earthly departure.

2. Dreyfus—Chess against the Machine

Having provided a symbolic analysis of Bergman's cinematic portrayal of *chess,* an important context has been established. What Bergman shows is that when human beings play chess, they play a game that challenges both their bodies and their minds. Indeed, Bergman succeeds in conveying deep existential insight by "arguing" that when Death takes human form to play chess, it is inevitable that he comes to place limits upon his supernatural powers.

On the basis of these general considerations, it will now be possible to effectively *highlight the specific dimensions of embodiment that Hubert Dreyfus's account of chess playing fails to register.* In order to do so, we move from symbolic issues concerning the human condition, temporality, and reflection, to the theme of what the human experiences of perception and action entail—particularly with respect to matters of temporality and reflection. The transition to this theme, however, requires preliminary consideration of Dreyfus's views on how computers approach the game of chess. These ruminations have appeared in numerous books and articles, and they form the basis of his interview with

Jim Lehrer after Garry Kasparov, World Champion from 1980 until 2000, lost to I.B.M.'s Deep Blue. According to Dreyfus, the computers that are currently programmed to "play" chess are not designed to understand its nuances. During matches they do not, properly speaking, engage in the human activity of play. This is because they do not "think" about what they or their opponents are doing; neither do they feel anxiety, elation, or any affect at all. Whereas humans find it difficult to avoid taking (what the philosopher Daniel Dennett calls) the "intentional stance" when they encounter machines that give the appearance of intelligent behavior, "chess-playing" computers do not impute deception to bluffing players or boldness to unorthodox moves. What makes "chess-playing" computers so successful, then, is that they are brute calculators who simplify the game into a "micro-world," i.e., a mathematical problem that can be formalized through extensive and efficacious calculations.[8]

Succinctly put, the success at chess displayed by computers lies in their ability to rapidly and consistently assign weighted numerical values to many facets of the game in order to render a quantifiable evaluation of a given position—an evaluation that will determine, algorithmically, which move is best to execute.[9] In this context, the mathematical technique, *alpha-beta pruning*, enables computers to limit the number of alternative moves that require analysis, and *heuristics*, which are basically "rules of thumb," determine which kinds of moves computers will examine and attempt first. Some computers, such as Deep Blue, are even programmed to compare—mostly through programmed input— how Grandmasters have assessed positions with their own computational determinations. In addition to this impressive feature, Deep Blue also uses *selective-search* to identify scenarios where a certain number of moves involve forced responses.

Given the pronounced disparity between the calculative powers

---

[8] David Suits suggests that it is probably incorrect to refer to computers as "solving" chess problems. The formalism involved, Suits claims, probably entails that computers treat chess like a "puzzle," i.e., like an entity for which there is no known solution to the whole thing, but only useful and less useful ways of addressing the well-defined parts.

[9] Such calculations include, but are not limited to: computing the value of the pieces found on both sides of the board, determining how much of the board each side manages, measuring how many squares a given piece can move to, *et cetera*.

that humans and computers display, Dreyfus claims that humans and computers clearly engage with chess in different ways. While Deep Blue could analyze an astonishing sixty billion board positions in the three minutes that players are allotted to select their moves, at best Grandmasters can only evaluate three board positions per second (Litch 2002, 91). And yet, despite being outmatched on a calculative scale, Kasparov somehow proved capable of defeating Deep Blue on more than one occasion!

Before clarifying what accounts for the disparity between how humans and computers engage with chess, one final contextual remark is in order—

It has already been noted that stark differences obtain when contrasting the mathematical capacities of human experts with digital computers. In addition to validating this observation, Dreyfus and his mathematician brother Stuart further claim that the mathematician's aptitude for problem-solving can actually hinder his or her acquisition of advanced chess skills. While *prima facie* this view might appear counter-intuitive, the proposition is actually reasonable. The assertion is that brute calculation works well, but only so long as it can occur at the speed of a digital computer:

> Stuart and his mathematical friends never got beyond the competent level. Students of math may predominate among chess enthusiasts, but a truck driver is as likely as a mathematician to be among the world's best players. Stuart says he is glad that his analytic approach to chess stymied his progress because it helped him to see there is more to skill than reasoning. (Dreyfus and Dreyfus 2004, 400)

2b. Chess Players: Skilled Humans

Given everything that has been noted in the previous section, the question to be asked is: How, then, do Grandmasters go about playing chess? According to Dreyfus, the fundamental point of contrast between humans and computers is that Grandmasters typically do not deliberate about strategy or tactics. Unlike computers, they simply perceive thepatterns that the chess pieces form as meaningful images and respond accordingly. Humans may not be naturally disposed to memorize and execute long lists of rules, including complex heuristics for playing successful chess games, but they do not need to be—not to excel at chess, and not even to engage in long-term planning in general.

Such immediate responsivity to chess patterns is possible, Dreyfus declares, because the human perceptual system can reliably and rapidly associate particular sensory inputs with behavioral responses that are appropriate, given the context and the agent's needs and desires. Simply put, our perceptual system may require external structures (e.g., writing instruments and surfaces) and culturally standardized forms of symbolic notation (e.g., numbers and letters) to process complex logical operations, but even when these tools for cognitive scaffolding aren't present, the human perceiver still remains primed for pattern recognition. Dreyfus speculates, for example, that Grandmasters can "directly discriminate perhaps hundreds of thousands of types of whole positions" (2005, 55).

In characterizing the *Gestalt* experience of Grandmasters as "direct," Dreyfus claims that in the blink of an eye they can visually identify chess patterns as relevant and "compelling." Such instantaneous perception is possible, he insists, because their perceptual acts are not slowed down by processes of conceptual and computational processing. Succinctly put, mediated acts, such as introspecting about each of the opponent's moves in order to mentally translate them into identifiable structures from which explicit inferences would be drawn, are too cognitively expensive. For example, it is inefficient for Grandmasters to perform the following two-part process: first, conceptualize the pattern of an opening sequence as a variation of the "Sicilian Defense," then deliberate about retaliatory options by evaluating from amongst the possibilities that internal memory stores as a list of potentially useful counter-sequences. Such formalized behavior may be necessary for players with moderate skills, but Grandmasters can get better and quicker results without performing a decision-tree style of analysis.

In order to ensure that Dreyfus's account does not appear unduly mystical—mostly because it posits such stark cognitive and perceptual differences between experts and novices—a brief comparison between Grandmasters and skilled athletes who compete in physically intense domains is in order. Instead of waiting for their brains to convert their visual fields into symbolic forms, Grandmasters discipline their hands to follow the same reflexive responses exhibited by athletes who enter into a motivational state that sports psychologists term the "zone."[10] In the zone, the primary comportment that players experience the world through is "flow." Flow is a state of focus that is plagued by neither boredom nor anxiety; it allows for such attuned concentration to occur that consciousness can transfix on tasks without being hindered by distracting considerations. Just as flow enables ball players of all stripes to successfully pass their games' sacred object to teammates without experiencing a reflexive moment in which they will their hands to open up from a tight grip to a pulsating release, so too do Grandmasters' hands respond to perceptual "affordances" by "spontaneously" moving chess pieces to their appropriate-looking locations.[11]

Again, such spontaneity is not random. Rather, it is a *normative* perceptual response in which the body is drawn to get a "maximal grip" on its environment (Dreyfus 2005, 57). Dreyfus considers such perception to be normative because it has "satisfaction conditions" that, to quote philosopher Sean Kelly, "say something about how the world *ought to be* for me to see it better" (Dreyfus 2005, 57). The correlative elation that comes from executing this type of direct *Gestalt*-motivated action is normative as well. Grandmasters and athletes do not typically become elated because arbitrary standards justify their performance as excellent. Instead, they experience positive affect because they excel at tasks that require them to push their skills to the limit, whatever their limitations have evolved to be.

To summarize—

From Dreyfus' perspective, Grandmasters do not, ordinarily, deliberate—in any formal and sequentially calculative sense—about what to do. They simply allow "intuition" to motivate their deeply attuned conduct, and by giving in to this disposition enter into a trance-like state—a feeling of harmony in which the chess player becomes impervious to distraction and perceives him or herself to be united with the objects of engagement.[12] As psycho-

---

[10] Dreyfus, appropriating the term from ecological psychologist J.J. Gibson, also refers to the perceptual inputs as the "ambient optic array" (2005, 54).

[11] "Affordance" is another term that Dreyfus appropriates from Gibson (2005, 56).

[12] Although Dreyfus claims that the Grandmaster's play is ordinarily bereft of deliberation and rule-following, he does acknowledge that there can be occasions in which "some sort of disturbance occurs"; in such instances, a

logist Mihaly Csikszentmihalyi, the theorist credited with coining the term and study of flow, notes: "The concentration is like breathing—you never think of it. The roof could fall in and, if it missed you, you would be unaware of it" (39).

To crystallize these points about immediate, direct, and focused responsivity, Dreyfus invites us to consider the case of "lightning chess":

> Grandmasters must make some of their moves as quickly as they can move their arms—less than a second a move—and yet they still play Master level games. When the Grandmaster is playing lightning chess, as far as he can tell, he is simply responding to the patterns on the board. At this speed he must depend entirely on perception and not at all on analysis and comparison of alternatives. (2005, 53)[13]

As further evidence of the fact that Grandmasters do not play chess by mentally performing intricate mental calculations, he and Stuart also offer up the following experimental "data":

> We've recently performed an experiment in which an international chess master, Julio Kaplan, had to add numbers at the rate of about one per second while playing five-second-a move chess against a slightly weaker but master-level player. Even with his analytical mind

---

skilled player may need to deliberate about why a particular move did not call for an immediate response, or why several responses were solicited "with equal pull" (2005, 57). Even the "rules of the game" (i.e., the particular moves each piece can make, the moves that count as cheating, *et cetera*), Dreyfus declares, are not normally internalized as policies that skilled players follow—either consciously or unconsciously (2005, 53). Instead, the rules that novices are typically taught through formal instruction become cognitively transformed; through the acquisition of experience-based skill, the rules come to be "experienced in the background as a limit on what appears as worth doing" (2005, 53). Skilled players, in this sense, display "sensitivity" to the rules of the game simply by virtue of experiencing a constrained perception of how to proceed during a given turn (Dreyfus 2005, 53).

[13] Ben Hale notes that given how "lightening chess" differs from tournament chess, it is debatable as to whether it deserves to be considered "chess" at all. This is a good question, one whose answer would bring us beyond the scope of this essay. For present purposes, it will be useful to follow the logic of the argument presented by the Dreyfus brothers in order to determine how persuasive their account of intuition is.

apparently jammed by adding numbers, Kaplan more
than held his own against the master in a series of
games. Deprived of the time necessary to see problems
or construct plans, Kaplan still produced fluid and coordinated play. (Dreyfus and Dreyfus 2004, 400)[14]

While I think that, in general, Dreyfus correctly identifies the
crucial differences that demarcate how Grandmasters and computers play chess, I also find his analysis of perception and action
to be hindered by questionable assumptions. To clarify what these
assumptions are and how they can be corrected, we turn now to
the issue of perception and embodiment.

3. Perception and Embodiment

When Dreyfus discusses the Grandmaster's ability to "intuitively"
perceive what move to make, he asserts that such a style of interaction is, in actuality, a common form of human engagement.
It is a way of "being-in-the-world" that all experts share, and its
scope extends from mundane to more rarified interactions: "This
holds true for such refined skills as chess, jazz improvisation, sport,
martial arts, etc., but equally for everyday skills such as cooking
dinner, crossing a busy street, carrying on a conversation, or just
getting around in the world" (Dreyfus 2005, 58). What, then, accounts for such a potent perceptual disposition? Dreyfus's answer

---

[14] In order to fully detail Dreyfus's account of Grandmaster perception, it
will be helpful to place that description within the broader context that he
intends for it to be understood within. To accomplish this, a review of his
famous account of skill acquisition is in order. Dreyfus first developed the
basis for his descriptive account of expertise with Stuart during the 1960's,
when hired by the RAND Corporation as a consultant to evaluate their work
on artificial intelligence. His research culminated in a 1967 paper "Alchemy
and Artificial Intelligence," and this, in turn, provided the basis for his famous
1972 book *What Computers Can't Do*. But it was not until their 1986 collaboration *Mind Over Machine: The Power of Human Intuition and Expertise
in the Era of the Computer* that the two brothers developed a model of skill
acquisition whose scope is, they claim, universal. According to the Dreyfuses,
they found a way to detail how the typical human learner acquires intellectual
as well as practical skills by apprenticing a standardized pedagogical experience that begins with formal instruction and proceeds developmentally. Each
educational stage that they identify is cumulative, and each marks a distinct
advancement upon what previously transpired. Ultimately, they claim that
the sequencing of five distinct phases from "novice" to "expert" is a rite of
passage that culminates in the profoundest of cognitive and affective transformations. For more on this, see Selinger and Crease 2003.

is *embodiment*—more specifically, "the nonconceptual embodied coping skills that we share with animals and infants" (2005, 47). Accordingly, he writes:

> We need to consider the possibility that embodied be-ings like us take as input energy from the physical uni-verse and process it in such a way as to open them to a world organized in terms of their needs, interests, de-sires, and bodily capacities without their *minds* need-ing to impose a meaning on a meaningless given. . . nor their *brains* converting the stimulus input into reflex responses. (Dreyfus 2005, 49)[15]

While it is obvious that in moving his or her hand, the Grand-master moves a *body part*, it is less clear from Dreyfus's writing why the perception underlying that physical movement is attrib-utable to a *robust notion of embodiment*. Recognizing this lacuna, Harry Collins, a prominent sociologist, has recently charged that Dreyfus's analysis conflates "embodiment" with "embrainment." Collins's point is that the human brain is the organ responsible for our perceptual activity—even "minimally embodied" beings with human brains could, presumably, still play chess.[16] Indeed, despite the heavy rhetorical emphasis on embodiment, Dreyfus's actual attempts to establish the *scientific plausibility* of his ac-count of perception and action emphasize, of all things, brain and brain-like activity! His main appeals are to Walter Freeman's "idea of the brain as a nonlinear dynamical system" and "the way simulated neural networks can be programmed to produce reliable responses" (Dreyfus 2005, 49, 54). Moreover, in order to bolster his account of chess playing with recent neuro-scientific evidence, Dreyfus notes that "recent brain imaging confirms that amateur and expert chess players use different parts of their brain" (2005, 53). In other words, Dreyfus declares that a shift from "left to

---

[15] Evocations of "embodiment" occur throughout Dreyfus's discussion of intuition and expertise. One of the dominant motifs that he appeals to is Merleau-Ponty's notion that our bodies are drawn to obtain a "maximal grip" on the environment (Dreyfus 2005, 57). He also associates the phenomenolo-gical notion of "coping" as expressed in both Heidegger and Merleau-Ponty as well as Gibsonian "affordances" as activities that "embodied" agents perform (2005, 53-54).

[16] For more on this, see Collins 1992, 1996, 2000, 2004a, 2004b, Selinger 2003, Selinger and Mix 2004.

right hemisphere processing" occurs during the transition from "detached analytical rule-following" to "an entirely different engaged, holistic mode of experience" (Dreyfus 2005, 52). And, in what appears to be the most explicitly contradictory of his remarks, Dreyfus actually depicts Kasparov as a "disembodied mind": "There are cases when the affordance is relative to the *disembodied* mind. To Kasparov, but not to a merely competent player, a specific situation affords a checkmate" (2005, 35). In one sense, it is understandable that Dreyfus would rely on distinctly brain-oriented rhetoric. From a naturalist perspective, the brain and the body form an inseparable biological system—to talk about embodiment is thus to talk about brain states and functions. Moreover, in putting matters this way, Dreyfus can posit that his phenomenological descriptions are compatible with solid scientific finds, including (although he does not reference it) the discussion of "thin-slicing" made popular by Malcolm Gladwell in his recent *Blink: The Power of Thinking Without Thinking*. For reasons like these, I do not find it at all objectionable to discuss how the brain's capacities relate to our perceptual experiences. What I object to are the *conceptual implications that follow from the specific manner in which Dreyfus discursively shifts from emphasizing embodiment to focusing upon brain activity*. To clarify this point, a discussion of why Dreyfus's depiction of Kasparov is untenable is in order.

On a superficial level, the notion of a "disembodied mind" is inapplicable because Kasparov might, like other famous chess players, have a strongly embodied pre-game regime—a regime that influences what state of mind he's capable of playing in. And even if he doesn't, other obvious bodily states, such as being hungry or tired, can also impact how his "mind" functions. Moreover, because Kasparov's mind is embodied in a being who is historically and culturally embedded, it is likely that, like other chess players, he would find it easier to compete on a black-and-white board equipped with traditionally shaped pieces. Qualitatively different designs—such as the one created by the Bauhaus School in which the pieces are designed to reflect how they move—would probably take some time to get used to. This is particularly the case if Kasparov's current *Gestalt* perceptions are wedded to specific geometric forms. Given a charitable interpretation, it is likely that Dreyfus would assent to the validity of these claims. However— On a deeper level—one which a charitable interpretation cannot remedy—we can say that Dreyfus's error arises because the ac-

count of perception he provides is too insular; it portrays Grandmasters as always focusing exclusively on the patterns made by
the chess pieces, as always being in the "zone" and being unen-
cumbered by other distractions. In a previous essay and Chapter
1 of this book, Robert Crease and I put the point this way:

> Dreyfus assumes that the body which acquires skill has
> no relevant biography, gender, race, or age. He does
> acknowledge that "cultural styles" affect how skills
> are learned, noting for example that differences exist
> between how American and Japanese mothers "handle"
> their babies. However, this notion of "cultural style"
> is not developed beyond unsubstantiated generalities,
> and assumes as well the insignificance of any biographic
> differences existing within individuals sharing a single
> culture. From Dreyfus's perspective, one develops the
> affective comportment and intuitive capacity of an ex-
> pert solely by immersion into a practice; the skill-
> acquiring body is assumed to be able, in principle at
> least, to become the locus of intuition without influ-
> ence by forces external to the practice in which one is
> apprenticed. (Selinger and Crease 2003, 260-261)

To clarify further how Dreyfus's notion of the unmarked body
distorts his account of both the nature of human perception and
chess playing, we need to remind ourselves that while Dreyfus
dissociates his discussion of intuition from stereotypes about how
females have been alleged to perceive the world, male supremacy is
one of the stereotypical images that chess evokes. In *Chess Bitch:
Women in the Ultimate Intellectual Sport*, Jennifer Shahade, the
2004 Women's Chess Champion, notes: "In the United States,
fewer than three percent of competitive adult-rated players are
women... In the worldwide ranking system of FIDE... the situation is more balanced. There, about six percent of active adult
players are female" (3). Shahade presents these statistics in order
to entertain subsequently the variety of biological, psychological,
and social theories that have been proffered as explanations of this
asymmetry. Those views and their sources can be summarized as
follows:

- Kasparov himself alleges that women have lower capacity to
  concentrate than men because of their "maternal impulses"
  (4).

- American Grandmaster and Freudian psychologist Reuben Fine contends that women are less inclined to be drawn to chess than men because patricide is the unconscious motive for playing (4).

- German youth champion Elizabeth Paethz insists that due to evolutionary differences dating back to the division of gendered labor in the Stone Age, women have less of a capacity than men to engage in prolonged periods of focused attention (8-9).

- Grandmaster Susan Polgar argues that since the modern division of gendered labor routinely provides women with fewer opportunities than men to engage in abstract reasoning, women are typically less adept at strategizing than men are (9).

In all likelihood, these types of claims are merely ideological convictions that present conjecture and anecdote in scientific-sounding jargon in order to generate the appearance of objectivity. But beliefs of all sorts can be causally efficacious, and someone who takes them to be true or likely to be true will act accordingly. Thus, insofar as chauvinist explanations such as these prevail and influence how some chess competitors view their rivals (and possibly themselves), and in so far as some chess trainers actually construct their regimen around the idea that menstruation effects how women, including highly ranked ones, are capable of playing, it is preposterous to assume, as Dreyfus does, that Grandmasters necessarily extinguish or repress all game-influencing prejudices about gender—and race as well—before commencing play.[17] Indeed, one of the striking stylistic dimensions of Shahade's account is that it portrays many of the female competitions as experiences of camaraderie; by contrast, she suggests that male competitors tend to develop acrimonious rivalries marked by explicit hostility.

Thus, contrary to how Dreyfus depicts the essentials of playing chess, body language—including sizing up an opponent (e.g., taking note of beads of sweat, furrowed brows, clenched hands, *et*

---

[17] Shahade, for example, notes how surprised she was to learn that one of her Russian trainers wanted to know her menstruation cycle in order that she could change her game playing accordingly. He even mentioned a computer program that had been developed to "determine how, on any given day, the menstrual cycle could affect play" (15).

cetera) to determine strengths, weaknesses, and potential proclivities—plays an important role in face-to-face chess conduct.[18] Deep Blue may not be programmed to care about how long a player takes to make a move or whether the opponent's lip curls in a certain way, but people cannot help but take these behaviors as direct and meaningful communicative signs. As a consequence, a rigorous phenomenological analysis of chess would not follow Dreyfus's lead and restrict its focus to the player's visual attention to the chess pieces. Instead, it would also attend to the concrete environment that chess play occurs in. With respect to matters of embodiment, that environment is, as Shahade notes, *erotically fraught*. Players assume a relatively stationary bodily position, and, within the context of this rooted comportment, find their gaze alternating between human and non-human focal points : "The sexual symbolism in chess is a rich topic. Chess is an intellectually intimate game in which two players sit for hours, both gazing at each other as well as the chess board" (70).

In light of chess being such a charged context, it is not surprising to learn that Shahade also makes the following remarks:

- *Her game play is directly influenced by the type of person she is playing against*: "My own motivation spikes when I play against men that I admire or find attractive. I find it fun to play against someone I like, and therefore I work harder at the board... I realized that I could also experience heightened concentration against women I admire" (71).

- *Some men claim to find their concentration limited because they are prone to thinking about sex while playing chess*: "According to the 2003 American Champion, Alexander Shabalov (...) most men, regardless of their strength, are thinking about sex for most of the game" (6).[19]

- *Due to the presence and power of the "male gaze," there is*

---

[18] Body language likely also plays an important role in online chess games. If one knows that an opponent is human, then certain moves that are perceived as "retreats" will also be ones that are perceived to be inspired by "fear." While fear is something that, to some degree, can be learned through stories, the most basic connection to it that allows its complexities and proclivities to be understood is, of course, a personally embodied experience.

[19] Shabalov actually informs Shahade that during most games, he thinks "about girls for about fifty to seventy-five percent of the time" (Shahade 2005, 6).

*an "invisible" but potent difference between how men and
women think and feel*: "At tournaments, women may find it
more difficult than men to completely lose themselves in the
game and reach a zen-like state of total focus. That women
are trained from a very early age to be constantly aware of
how they appear may explain this... Such an extra layer of
self-consciousness makes it hard to experience life directly
or to feel pure freedom" (156).

This type of testimonial evidence surely does not present us
with sufficient reason to believe that culturally embodied features
(such as prejudice) will, of necessity, causally influence all—or, indeed, any of—the *Gestalts* that Grandmasters perceive. Indeed, it
is entirely possible that prejudiced (or amorous) people are able to
suspend their biases when playing chess. Bigots, in other words,
do not need to express bigotry during every moment they are conscious; they can still be remarkable chess players. What I have
been trying to show, therefore, is that Dreyfus's account of embodied chess-playing lack sufficient ontological rigor. Dreyfus's
philosophy suffers precisely because he pursues a descriptive orientation that is hermetically sealed; a more hermeneutic—which
is to say, interpretative—approach would be open, from the start,
to considering the question of how eros and prejudice might influence the ways that embodied Grandmaster's think, perceive, and
act.

Conclusion

What, then, are we to make of Dreyfus's account of Grandmasters in light of observations such as the ones Shahade makes?
After all, the concrete details she provides ultimately reinforce the
more symbolic and general account of embodied chess playing that
Bergmann puts forth. Both Shahade and Bergman agree that as
an embodied activity, chess cannot offer complete transcendence—
perceptual or otherwise. To return to the title of this chapter, it
can be said that *if computers could play chess like human Grandmasters, they would need to do more than become better at pattern
recognition. To play like humans, even the most skilled of us, they
would need to be capable of perceiving their opponents in erotically,
culturally, and biographically charged ways.*

Dreyfus could, of course, retort that because his account of chess
playing focuses on Grandmasters, it is immune to the kinds of
evidence and concerns that have been articulated thus far. From
Dreyfus's perspective, Grandmasters and lesser skilled chess play-

ers simply relate to the game in wholly different ways.

Before responding to this possible attempt at a refutation, one final contextual remark is in order. What needs to be emphasized is that Dreyfus is not interested in how Grandmasters perceive chess pieces simply because he wants to understand the intricacies of human perception. Rather, his philosophical intervention is oriented towards more explicitly normative concerns. What Dreyfus wants to do is establish criteria that demarcate *human judgment* from *computerized decision-making*.

Although Dreyfus does not use the term, he clearly depicts the Grandmaster as a *hero*. Indeed, he goes so far as to represent the Grandmaster as a *symbol of perceptual excellence* that the computational conception of mind has "endangered" to the threshold of extinction. He and Stuart thus contend that the pervasive misunderstanding of the type of intuition routinely displayed by Grandmasters has already contributed to several negative states of affairs (2004). Here are some examples of this misunderstanding in action:

- Diverse fields, including diagnostic medical domains, are beginning to use so-called "expert" computer systems. Even though intuitive human experts can generate more accurate results than such systems, zealous advocates still characterize this technology as cutting-edge.[20]

- Computer applications are being brought into the classroom to perform tasks ranging from tutoring to distance learning. Zealous advocates characterize these machines as indispensable even though such devices cannot replicate the quality of face-to-face, intuitive human pedagogical excellence.

- "Scientifically sound" methods of business entrepreneurship are currently being taught that fail to prepare business students to respond intuitively to contextual considerations. Despite this deficiency, zealous advocates represent these techniques as thorough.

- Extremely expensive (and tax-payer funded) artificial intelligence projects are sold to the U.S. Department of Defense

---

[20] Timothy Engström suggests that in order to clarify the differences that distinguish medical diagnoses made by computers from ones made by physicians, it would be necessary to distinguish their respective orientations towards temporality.

that cannot possibly yield the results that their zealous advocates promise. Such projects run the additional risk of adversely affecting how the civilian sector adapts to similar technology; automated systems can be slated to replace intuitive contributions.

It is precisely because Dreyfus wants to combat these scenarios philosophically—as well as other cases in which computationally instantiated mathematical algorithms threaten to guide crucial decisions that are "best left to human judgment"—that he offers up his analysis of how Grandmasters play. His narrative, in other words, is constructed in the spirit of *consciousness raising*. It is intended to produce the *inspirational effect* of reorienting how his readers treat their *core relationships to technology*.

Although it is commendable that Dreyfus wants to provide this service, the problem is that he sets the demarcation bar too high. While in *some cases* Grandmasters probably are able to play chess in the manner that Dreyfus depicts, he goes too far in assuming that *every time* they play Grandmaster style chess their relation to the game is always immune to the forces of *eros* and *prejudice*. If Grandmasters could consistently play in such a pristine and pure way, they would not be human. Instead, they would be relating to other people through the mediation of the chess pieces and board in an affect-inhibited manner that applies more to computerized deliberation than intersubjective human engagement. To admit this point, however, Dreyfus would have to concede that human intuition is subject to greater fallibility than he represents. One promising way to revise this type of inquiry would be to revisit classic matches such as the one between Kasparov and Deep Blue and examine how both competitors were coached in between games. Doing so might reveal that far from merely being a contest of "human vs. machine," it was a hybrid competition between human and non-human agents—both of which required external input (computer and human generated) to compensate for both human and computational limitations. Such investigation would also focus upon whether Kasparov's losses to Deep Blue are more directly attributable to his perceptions of what it meant to play against a computerized opponent (and from the inability of his support team to adequately deal with these perceptions) than any other failing.

Bibliography

Allen, W. 1978. Death Knocks. In: W. Allen, *Getting Even*,

pp. 26-30. New York: Vintage. Avni, A. 2002. *Practical Chess Psychology*. London: Batsford.

Collins, H.M. 1992. Dreyfus, Forms of Life, and a Simple Test For Machine Intelligence. *Social Studies of Science* 22: 726-739.

Collins, H.M. 1996. Embedded or Embodied? A Review of Hubert Dreyfus' *What Computers Still Can't Do*. *Artificial Intelligence* 80: 99–117.

Collins, H.M. 2000. Four Kinds of Knowledge, Two (Or Maybe Three) Kinds of Embodiment, and the Question of Artificial Intelligence. In: M. Wrathall and J. Malpas, (eds.) *Heidegger, Coping and Cognitive Science: Essays in Honor of Hubert L. Dreyfus*, vol.2, pp. 179-195. Cambridge: MIT Press.

Collins, H.M. and Evans, R. 2002. The Third Wave of Science Studies: Studies of Expertise and Experience. *Social Studies of Science* 32: 235-296.

Collins, H.M. 2004a. Interactional Expertise as a Third Kind of Knowledge. *Phenomenology and Cognitive Science* 3 (2): 125-143.

Collins, H.M. 2004b. The Trouble With Madeleine. *Phenomenology and Cognitive Science* 3 (2): 165-170.

Csikszentmihalyi, M. 2000. *Beyond Boredom and Anxiety: Experiencing Flow in Work and Play*. San Francisco: Jossey-Bass.

Dennett, D. 1989. *The Intentional Stance*. Massachusetts: MIT Press.

Dreyfus, H. 1992a. *What Computers Still Can't Do*. Massachusetts: MIT Press.

Dreyfus, H. 1992b. Response to Collins, *Artificial Experts*. *Social Studies of Science* 22: 717–26.

Dreyfus, H. 2000. Response to Collins. In: M. Wrathall and J. Malpas, (eds.) *Heidegger, Coping and Cognitive Science: Essays in Honor of Hubert L. Dreyfus*, vol.2, pp. 345-346. Cambridge: MIT Press.

Dreyfus, H. 2005. Overcoming the Myth of the Mental: How Philosophers Can Profit from the Phenomenology of Everyday Expertise. *Proceedings and Addresses of the American Philosophical Association* 79 (2): 47-65.

Dreyfus, H. 2001. *On the Internet*. New York: Routledge.

Dreyfus, H. and Dreyfus, S. 1986. *Mind Over Machine: The Power of Human Intuition and Expertise in the Era of the Computer*. New York: Free Press.

Dreyfus, H. and Dreyfus, S. 2004. Why Computers May Never Think Like People. In: D. Kaplan (ed.) *Readings in the Philosophy of Technology*, pp. 397-413. New York: Roman & Littlefield Publishers.

Gibson, J.J. 1979. *The Ecological Approach to Visual Perception*. Boston: Houghton Mifflin.

Gladwell, M. 2005. *Blink: The Power of Thinking Without Thinking*. New York: Little, Brown, and Company.

Heidegger, M. 1992. *The Fundamental Concepts of Metaphysics*. Bloomington: Indiana University Press.

Heidegger, M. 1993. Letter on Humanism. In: D. Krell (ed.) *Basic Writing*, pp. 217-265. San Francisco: Harper.

Ihde, D. and Selinger, E. 2004. Merleau-Ponty and Epistemology Engines. *Human Studies* 27 (4): 361-376.

Litch, M. 2002. *Philosophy Through Film*. New York: Routledge.

Selinger, E. 2003. The Necessity of Embodiment: The Dreyfus-Collins Debate. *Philosophy Today* 57 (3): 266-279.

Selinger, E. and Crease, R. 2002. Dreyfus on Expertise: The Limits of Phenomenological Analysis. *Continental Philosophy Review* 35: 245-279.

Selinger, E. and Mix, J. 2004. On Interactional Expertise: Pragmatic and Ontological Considerations. *Phenomenology and the Cognitive Sciences* 3, (2): 145-163.

Shahade, J. 2005. *Chess Bitch: Women in the Ultimate Intellectual Sport*. Los Angeles: Siles Press.

Wolff, P. 2005. *The Complete Idiot's Guide to Chess*. New York: Penguin Group.

Part II
Interactional Expertise

3

Interactional Expertise and Embodiment

Introduction

In an extensive body of work, Harry Collins applies sociological tools to a problem that he calls "linguistic socialization."[1] Because complete linguistic socialization is said to lead to fluency in a language according to the standards of a given discursive community, it is central to debates about *artificial intelligence* (AI) and *expertise*. While Collins is not the only theorist to contribute to these issues, he pursues a sustained research program on how much knowledge *language alone* can convey.

With respect to artificial intelligence, debate exists over whether computers will ever be able to pass the Turing Test. Controversy thus abounds as to whether—and, for some, when—machines will come to have natural language conversations that humans would count as intelligent. As Collins contends, for computers to succeed they will have to acquire not only information, but also tacit knowledge. Without tacit knowledge, semantic and referential problems will be insurmountable. (Collins 1990)

Collins's concept of "interactional expertise" stipulates that one can acquire all of the linguistic understanding of a domain by immersing oneself in the language of the domain and without actually engaging in its practices (2004a). It thus gives a positive answer to the question of how much understanding sociologists, activists, and journalists can obtain about a field through the medium of conversation alone. Not everyone agrees that it is possible

[1] Several people helped me think through the issues discussed here. In particular, I would like to thank Harry Collins, John Mix, Trevor Pinch, and Jack Sanders for their openness and availability. I am also grateful that the following people provided stimulating conversations and thoughtful editorial suggestions: Muhammad Aurangzeb Ahmad, Robb Eason, Timothy Engström, Patrick Grim, Robert Rosenberger, David Suits, Craig Selinger, and Noreen Selinger.

to learn to talk competently about aspects of a technical field (e.g., pass on technical information, assume a sound devil's advocate position on a technical matter, and even make authoritative judgments on a peer review committee) solely by immersing oneself in the talk of the field's "contributory experts"—that is, with people who are physically immersed in the field and who are capable of advancing it.[2] The purpose of this chapter is to demonstrate that while Collins furthers philosophical and sociological discussions on how knowledge and language relate, his overall position remains predicated upon a misunderstanding of the phenomenology of embodiment. *By omitting developmental consideration of how humans develop linguistic competence or skill itself, Collins misrepresents how knowledge is acquired as well as what kinds of people expert knowers truly are.* On the issue of artificial intelligence, Collins misinterprets the empirical evidence that he introduces for the purpose of supporting his anti-phenomenological claims.

Before establishing these critical points, it will be useful to begin with an overview of Collins's position. I will detail how Collins characterizes his outlook. For purposes of exposition it is useful to examine how Collins contrasts his position with the phenomenological perspective exemplified by Hubert Dreyfus. I will then present a phenomenological position on embodiment and explain why several of Collins's claims are invalid.[3]

To avoid potential confusion for the reader, some caveats are in order—Although I will be discussing views on the Turing Test throughout the chapter, my aim is not to speculate on whether or not machines will ever pass a generalized test of this kind – that is a test that is not domain-restricted. Instead, the analysis that follows is deflationary. I intend to demonstrate that Collins does not illuminate the topics of artificial intelligence and interactional expertise in the manner that he thinks he does. While I will

---

[2] Even Collins, at an earlier period of his career, espoused the contrary position. Not taking into account how much a spy could learn from talking with natives of a given city, Collins claimed that if a native asked the right questions to an infiltrating spy, the spy would be revealed as a fraud who originated elsewhere. The switch of position is discussed in Collins and Evans, 2007.

[3] Although I will be drawing upon previous critiques, the arguments that follow are intended to further previous discussions (Selinger 2003; Selinger and Mix 2004).

address how humans typically develop their cognitive architecture, I will not discuss whether it is possible for that architecture (or some analogue to it) to be instantiated in a computer. Indeed, I am agnostic about whether a 'minimally embodied computer' (Collins's term), could one day produce behavior indistinguishable from successful linguistic socialization. My objective is simply to establish the reasons why the data Collins has gathered does not bear on this question either.

Put in more positive terms, by undermining Collins's examples, I will reintroduce the possibility that human embodiment may be important not only for understanding how humans actually accomplish linguistic socialization but also for figuring out how we might design artificial intelligences. *This is not to say that human embodiment is necessary for such socialization* but rather that it is worth more consideration than Collins gives it.

1.   Collins on Computers and Linguistic Socialization: The Basic Position

Inquiry into whether computers can be intelligent runs the risk of essentializing the notion of intelligence. This is because there is no single metric for intelligence. Not only is human performance routinely assessed according to different scales that measure different types of intelligences, but it may be reasonable to deem computers intelligent (in some respects) by judging them according to computational standards of performance, or else by assessing the kinds of results they can produce when they collaborate with human beings. Nevertheless, an optimally designed Turing Test demonstrates how much of an understanding of human social life, as revealed by linguistic competence, computers can imitate, or, perhaps, acquire. The question that Collins's notion of interactional expertise raises for the phenomenologist is whether computers will have to possess human or human-like bodies to acquire enough linguistic competence to pass the Turing Test.

Collins says that artificial intelligences do not need bodies even they he also says that they will not be constructed in the foreseeable future because we do not know how to socialize them. Linguistic socialization, he argues, is not crucially dependent on the body. From his perspective, phenomenologists' claims about the significance of embodiment are overstatements (Collins 1990, 1992, 1996, 2000, 2004a, 2004b; Collins and Kusch 1999; Collins *et al.* 2006). Collins says that, given the means to communicate and appropriate programming, an immobile box could acquire the "common sense" required to discuss anything pertaining to a typical human form-of-life and pass the Turing Test. It would some-how have to be connected to the social "form-of-life" in which the discourse was spoken but this would not require more than a minimal body. Collins agrees that we do not have any idea what the proper programming would consist of nor do we know how to create the means of communication. The crucial point is that the shape or mobility of the body would not be the limiting factor.

To clarify this proposition, Collins sets out two interrelated views on embodiment: the *"social embodiment thesis"* and the *"minimal embodiment thesis"* (2000). According to the social em-bodiment thesis, any particular language that develops can only be completely understood in terms of the bodies of the agents in that culture. For example, human culinary practices cannot be exhaustively analyzed without discussing biology (e.g., the need for organic life to acquire nutrition) and human physiology (e.g., the contours and sensory dimensions of our hands, mouths, and tongues, our olfactory proclivities, etc.). While this seems to argue for the importance of the body, in Collins's treatment the body is important only at the collective level. Turning to individuals, he claims that it is possible for someone who lacks the type of embodiment that is prevalent in a given society (and which has given rise to the prevalent language of that society) to be linguist-ically socialized as a member of that society. This is the minimal embodiment thesis—minimal embodiment being just enough em-bodiment to engage in successful conversation.

Collins concedes, then, that linguistic socialization cannot oc-cur in the absence of a being receiving aural inputs and emit-ting linguistic outputs. A socializable agent must possess ears (or their equivalent), a larynx (or its equivalent), and enough of the human-like brain —or computational analogue of it—to de-code what comes through the ears and reproduce the outputs of the human-like larynx. Even an immobile box would need these features to become linguistically socialized into our world.

Given the diverse forms of embodiment that can fulfill such minimal conditions, Collins does not limit his views on linguistic socialization to the future potential of intelligent computers. "In principle," Collins writes, "if one could find a lion cub that had the potential to have conversations, one could bring it up in human society to speak about chairs as we do in spite of its funny legs". (1996, 104)

2. The Collins-Dreyfus Debate: Collins's Perspective

Collins characterizes his account of linguistic socialization as a position that stands in contrast to the phenomenological critique of artificial intelligence. He takes Hubert Dreyfus as his example of this phenomenological critique as represented by works such as *What Computers Can't Do*. According to Collins's interpretation, Dreyfus advances the following claim: Until computers become embodied like humans, they will, in principle, be unable to make sense of the perspectives that humans take when perceiving, acting, and judging; lacking human perspectives, they will be incapable of passing the Turing Test.

In contrast, Collins believes that (the right kind of) talking computers and the (right kind) of talking lions would be examples of minimally embodied beings who are, in principle, capable of passing the Turing Test. No human topic would be off-limits to them, not even discussions of intimate human experiences, such as love.[4] Collins believes, furthermore, that it is possible to refute Dreyfus by reference to already existing *empirical data* concerning successful instances of minimally embodied persons becoming linguistically socialized. In this context, he describes three cases—a severely handicapped woman named Madeleine, colorblind people, and a sociologist (himself) trying to master the discourse of gravitational wave physics.

2a. The Case of Madeleine

In the *Man who Mistook His Wife for a Hat*, Oliver Sacks describes one of his patients, Madeleine, as a "congenitally blind woman with cerebral palsy," who, for most of her life, experienced her hands as "useless godforsaken lumps" (59). Douglas Lenat, a computer scientist, appropriates this description as an empirical counter-example that putatively disproves Dreyfus's position on the cognitive importance of human embodiment. From Lenat's perspective, because it was possible for Madeleine to acquire common sense knowledge from books that were read to her, she is living proof that human embodiment is not decisive for learning natural language.[5] Indeed, despite her severe limitations, Sacks

---

[4] Collins and Kusch (1999) discuss what is required to write love-letters and conclude that a distinction needs to be drawn between knowing the emotion of love first-hand and being able to discuss love as a consequence of linguistic socialization into a community where agents frequently claim to have experienced love. See also Collins and Evans 2007.

[5] When Madeleine could not control her hands, it was not possible for her to read Braille.

describes Madeleine as an engaging conversationalist.

Dreyfus dismisses Lenat's argument on the grounds that Lenat's disembodied characterization of Madeleine is a distortion of the type of person she is. Dreyfus insists that although Madeleine is disabled, she still shares many of core features of phenomenological embodiment that able-bodied people experience: "She has feelings, both physical and emotional, and a body that has an inside and an outside and can be moved around in the world. Thus, she can empathize with others and to some extent share the skillful way they encounter the world". (Dreyfus 1992, xx)

Collins sides with Lenat on this issue—though not on the matter of whether Lenat's favored method of instilling language in a computer would work—and he extends the conversation by providing reasons why Dreyfus's reply should be understood as an inconsistent position. According to Collins, the only thing that Dreyfus's response establishes is that Dreyfus himself actually fails to grasp what embodiment is—that is, whether embodiment is something "physical" or whether it is something "conceptual":

> But under this argument [i.e., the Madeleine example] a body is not so much a physical thing as a conceptual structure. If you can have a body unlike the norm and as unable to use tools, chairs, blind persons' canes and so forth as Madeleine's, yet still have common sense knowledge, then something like today's computers... might also acquire common sense given the right programming. It is no longer necessary for machines to move around the world like robots in order to be aware of their situation and exhibit "intelligence". (1996, 104)

As Collins sees it, not only is Madeleine mostly a "brain" endowed with "some sensory inputs," but, and also contrary to Dreyfus, either no body or not much of a body is required in order to experience empathy and imagination (2000, 188; 2004, 125). Given this depiction, Collins treats the case of Madeleine as empirical proof that a minimally embodied being can become socialized by means of language alone.

2b. The Cases of Colorblindness and Gravitational Wave Physics

Because Sacks's account comes to us as a second-hand testimony for a popular audience, Collins then tries to gather first-hand empirical evidence from his own experiments. Since he cannot find any more "Madeleines," he tries to make the point of principle in the case of people with lesser deficiencies. In one case, Collins attempts to determine if colorblind people are capable of passing a Turing Test in which conversation focuses on color discourse. According to his recently published study, it turns out that they can (Collins *et al.* 2006).

The results that Collins obtains seem to confirm his views that bodily ability is not a necessary condition for linguistic ability; the colorblind can talk about color as fluently as color perceivers. On the face of it, it seems that colorblind people acquire their ability to speak as though there were color in their world from talk within the society of color perceiving people; they do not seem to learn much from direct experience with color sensations.

Since this study revolves around human subjects, it is worth putting emphasis upon the parallels that Collins intends to draw between it and the prospects for artificial intelligence. According to Collins, if colorblind people can pass a Turing Test on color discourse even though their visual systems do not process all of the colors that they can talk about, then additional empirical evidence has been obtained that establishes the possibility that, someday, linguistically socialized computers might be capable of passing a Turing Test on color discourse (or some other perceptual discourse), even if these computers cannot perceive color (or experience some other perception) in the manner that color-perceiving people (or some other group of human perceivers) can.

Collins's third case arises from the fact that despite being trained as a sociologist (and not a natural scientist), he successfully developed such an extensive understanding of gravitational wave physics that, under Turing Test conditions, practicing gravitational wave physicists were unable to distinguish between him and their colleague practicing gravitational wave physicists (Collins *et al.*, 2006). Collins takes this discursive success as confirmation that he has acquired considerable interactional expertise in gravitational wave physics. In other words, Collins contends that just as colorblind people learn to converse about colors (in the absence of being immersed in the practice of color perception), and just as Madeleine learned to converse about a range of human affairs (in the absence of being immersed in their practices), so too did he learn to converse about gravitational wave physics (in the absence of being immersed in the practice of gravitational wave physics research). All of these accomplishments are possible, according to Collins, because linguistic socialization alone can convey complete linguistic fluency.

This view on linguistic socialization appears, then, to be a radical departure from the philosophical position represented by Dreyfus. Dreyfus argues that one cannot become linguistically socialized into an expert practice without being a contributing expert practitioner oneself. For Dreyfus, in order to be capable of saying everything that can be said about a phenomenon, domain, or experience, one must immerse oneself fully in the corresponding physical activities. Using the practice of surgery as a paradigm case, Dreyfus writes:

> There is surely a way that two expert surgeons can use language to point out important aspects of a situation to each other during a delicate operation. Such authentic language would presuppose a shared background understanding and only make sense to experts currently involved in a shared situation. (2000a, 308)

On the Dreyfusian view, only a surgeon can have the appropriate background understanding to take part in the full range of surgical discourse—a discourse that is said to include special "authentic" conversational terms and norms. Even though Dreyfus's quote refers to a cooperatively performed operation and not to a Turing Test, use of the qualifier "only" places austere restrictions on the type of people who qualify as candidates for linguistic socialization. To claim that "only ... [the] experts currently involved in a shared situation" can appreciate the full range of contributory discourse, is to restrict the class of "authentic language" users to the set of people who have a "shared background understanding"—an understanding that, in the Dreyfusian view, can only be acquired if there is physical activity in a "shared situation." Given these restrictions, medical journalists, medical sociologists, computers programmed with social medical software, and talking lions that lack opposable human thumbs would be unable to acquire the full linguistic proficiency in surgical discourse of practicing surgeons.[6]

---

[6] If I am attributing too strong of a position to Dreyfus, it is only because his rhetoric is suggestive of the argument that I am making. Dreyfus clearly argues that experts cannot reproduce all of their perceptions and judgments in propositional form. In this sense, both Dreyfus and Collins agree that experts cannot pass on everything they know through discourse. But Dreyfus also argues that experts typically develop their skills by undergoing a five-step

On this last issue, the empirical evidence on interactional expertise that Collins introduces suggests that his position on what someone can say in the absence of direct experience is tenable. Based on the colorblind and gravitational wave physics experiments, it does not appear to be necessary to have first-person experience of a phenomenon in order to be capable of saying everything that humans can say about that phenomenon. Of course, one could argue that it is a mistake to draw inferences about experience in general based solely on the limited domains that Collins examines. But since Collins shifts the issue from a speculative to an empirical matter, the burden of proof falls on those who hold a Dreyfus-like position.

Having agreed with Collins on this aspect of conversational practice, I find the stronger claims for the power of linguistic socialization made by Collins untenable. By adhering to the minimal embodiment thesis, Collins fails to make crucial distinctions between experience and embodiment. I claim that it is largely because of embodied learning and embodied perception that humans can learn to speak about things they have not experienced and that the fluency acquired by Madeleine, by the colorblind and by Collins in gravitational wave physics depends crucially on embodiment.

3. Phenomenological Embodiment

Two interconnected reasons can be identified that explain why Collins's account of linguistic socialization is flawed.

1. By misunderstanding the phenomenological position on embodiment, Collins misrepresents the views of an established tradition.

2. By failing to provide a comprehensive description of bodily activity in the relevant empirical cases, Collins misinterprets the extant data on linguistic socialization.

If Collins better understood phenomenology, then—

- He would be less inclined to misdescribe how Madeleine and colorblind people become skillful conversationalists.

---

experiential process that runs from "novice" to "expert." In this context, he suggests that if someone does undertake all five of these stages, he or she will be unable to look at phenomenon as experts do.

- He would be less inclined to draw analogical connections between Madeleine, the colorblind, himself, and the potential computers of the future.

In order to establish these points, it will be helpful to proceed by considering some of the reasons why Collins might fail to grasp the type of analysis that phenomenologists like Dreyfus provide. Collins's point is not dissimilar to that of Dutch philosopher Philip Brey who writes:

> Human beings can have limbs and organs amputated or paralyzed and still not lose their ability to engage in abstract thought, and it is at least theoretically possible that, as sometimes depicted in science fiction stories, a brain could be removed from a body and kept in laboratory conditions while still retaining the ability to think. (2001, 51)

Moreover, if we add technological considerations to this list of examples, Collins's case appears to be strengthened yet further. Consider the following example discussed in Nicolas Humphrey's *Seeing Red*:

> An apparatus, called vOICe, has recently been developed for helping blind people to see using their ears rather than their eyes. The subject wears a helmet with a video camera mounted on it, coupled to a light-to-sound translation program, with some headphones to receive the sound images. The device has the potential to map visual scenes to "soundscapes" in an analog way. Future versions will likely code color as a continuous extra dimension of the soundscape. However, as of now, when it comes to "seeing color," the device takes a short-cut and *says* the word RED! As the user's manual explains, when you activate a color identification button, "the talking color probe speaks the color detected in the center of the camera view. Now you know whether the apple you are about to eat is yellow, green or red, and you can check the dominant colors of your clothing". (2006, 23)

Given these considerations, Collins might appear justified in pressing his question: What are the minimum body-parts that someone must have to become linguistically socialized? Similarly, Collins might also appear justified in being perplexed over Dreyfus's claim that having a human "front and back" helped Madeleine to become linguistically socialized. After all, giving a computer a human front and back will not turn a linguistically unsocializable computer into one that can be socialized.

How might a philosopher like Dreyfus respond to these charges? One reply would entail demonstrating that Collins misinterprets the data on Madeleine, the colorblind, and his own gravitational wave physics experiences. Such reinterpretation would emphasize the first-person perspective of the "lived body," and it would center on the two core issues that Brey associates with Dreyfus's phenomenology (2001, 50-51).[7]

1. Human perception and sensorimotor intelligence are "not localized in the brain," but instead are distributed through "a complex feedback system that comprises the nervous system, senses, the glands, and the muscles."

2. Humans develop abstract intelligence through sensorimotor activities, including activities that are conducted during the periods ranging from infancy through childhood. In order to comprehend these activities fully, the emotional and motivational processes that direct human action must be taken into account.

In what follows, I will provide my own phenomenological interpretation of the data that Collins takes to be relevant to his account. Although I do not wish to claim that my analysis is equivalent to the perspective that Dreyfus himself would advance, were he to attend closely to these matters, I nevertheless will emphasize the two considerations that Brey highlights.

One of the main points that I intend to establish is that while losing, or being born without, certain physical abilities clearly (and sometimes dramatically) influences how an embodied human agent relates to the world, situations of bodily debilitation, bodily depravation, and bodily diminishment tend to be situations in which more perceptual activity of an analogically related sort

---

[7] I am restricting my focus to dimensions of embodiment that are relevant to the Collins-Dreyfus debate. A comprehensive analysis of Dreyfus's views of embodiment would also need to attend to more overtly existential issues, such as "anxiety," "vulnerability," "commitment," "risk," "style," etc.

occurs than Collins acknowledges.

Since I do not have direct access to the data at issue, I will be reconstructing situations pertaining to perceptual experience and the acquisition of skill. Admittedly, such reconstruction is not scientific. Because it is not scientific, Collins may be inclined to dismiss my claims as *ad hoc* hypotheses that are postulated to preserve phenomenological assumptions. The burden of proof, however, should be seen the other way round. The reconstructions that I will be providing accord with common developmental experiences. If I make claims that Collins disagrees with, it is also his responsibility to demonstrate where they fall short. He would finally need to present a developmental account of language acquisition, and, possibly, concept-formation. As the next section will clarify, given the possible acts of compensation that Madeleine and colorblind people engage in, Collins might even benefit from consulting recent debates about "sensory substitution," i.e., "the possibility of substituting one kind of sensory input for another" (Humphrey 2006, 54).

4. A Phenomenological Interpretation of Collins's Data

4a. Madeleine Revised

Let's begin with Madeline. Given her disabilities, she is a harder case than the others to discuss in embodied terms. By comparison, the colorblind and gravitational wave physics experiments are minor examples. From a phenomenological perspective, some of the relevant aspects of Madeleine's situation can be reconstructed as follows.

Madeleine's basic perceptual experiences are *distributed throughout neural-somatic networks*. Since Madeleine is blind and has cerebral palsy, her eyes, hands, and legs may contribute less to her perceptual experiences than the eyes, hands, and legs of able-bodied people. In this respect, Madeleine is differently embodied than able-bodied people. Nevertheless, it is inappropriate to take up the standard of the minimal embodiment thesis and claim that Madeleine receives "inputs" from the world and communicates "outputs" back to it simply by possessing a human brain connected to a few discrete body parts. One way to see why Madeleine's experiences are irreducible to the conditions stipulated in the minimal embodiment thesis is to phenomenologically revisit the experience of empathy.

Whereas Collins claims that the human experience of empathy occurs in the brain, a phenomenologist would see the human experience of empathy as *distributed throughout the whole human body*—that is, as dispersed through coordinated *neural and somatic activity*. The way that we experience empathy is not unique; empathy has a complex and distributed structure because *affective human experiences in general* are like this. In "Empathy and Consciousness" Evan Thompson writes:

> Affect has numerous dimensions that bind together virtually every aspect of the organism—the psychosomatic networks of the nervous system, immune system, and endocrine system; physiological changes in the autonomic nervous system; the limbic system, and the superior cortex; facial motor changes and global differential motor readiness for withdrawal; subjective experiences along a pleasure-displeasure valence; social signaling and coupling; and conscious evaluation and assessment. (2001, 4)

Thompson's account of affect accords with first-person human experience and findings in biology and the neurosciences. In this respect, speculation is not required to establish that Madeleine's experiences of affect are experiences that she feels throughout her body. Madeleine may be handicapped, but she does not live in a human body that is so physically dissimilar to other human bodies that she experiences affect in an anomalous, non-human way. If Collins believes otherwise, he owes us an account that justifies this conviction. At present, Collins merely asserts that empathy is an "embrained" human experience; no supporting evidence is provided.

To clarify this point further, it will be useful to use some simple reconstructive phenomenology. In this context, a *developmental* issue needs to be clarified. Since empathy entails identifying with other people's feelings and motives, it is important to know how Madeleine experiences her own feelings and motives. Only the outline of an answer to this question needs to be provided to show where Collins errs; debates about the "simulation theory of mind" and the "theory theory of mind" need not be entertained.

For starters, it seems clear that if Madeleine is afraid, she does not relegate this experience to brain activity alone. Depending on the severity of the fright, Madeleine would have first-person awareness of her respiratory activity and heart rate accelerating as involuntary responses, and she would feel her lips quiver.[8] She might not feel her hands tremble in the same way as able-bodied people do in similar situations, but even if this is so, it simply means that her phenomenological experience of fear occurs through *a less replete neural-somatic* network than might be operative in others. Again, this is a comparative assessment in which different humans are juxtaposed; Madeleine remains more embodied than the parameters of the minimal embodiment thesis suggest. To elaborate further, when Madeleine is angry, she likely has first-person awareness of her facial muscles shifting, even though able-bodied people might have the same experience and also be aware of their posture shifting. Here, too, the only reductive comment about Madeleine's embodiment that can be made is comparative in nature.

Ultimately, Madeleine has such a rich experience of affect because she shares in our *human evolutionary history*. Within these parameters, the human organism has developed rapid means for responding to real and potential predators. Consequently, Madeleine and the rest of us become aware of affect, in part, through involuntary biological processes that are distributed throughout our bodies. This is not to say that social input is an insignificant index for people becoming aware of their emotions. To be sure, we may not be fully aware that we are afraid until someone asks: Why are you so nervous? But, again, conversational input is merely one component of how people can learn of their emotional states. The critical points are: (1) human brains alone are incapable of perceiving anything; and (2) human brains that are augmented only by the minimal input/output receptors detailed in the minimal embodiment thesis would find it difficult, if not impossible, to perceive an affect like fear in time to respond to it appropriately. John Mix and I articulated this outlook in an earlier article when we wrote:

> Depending on the intensity of the experience, fear can affect the human body in many different ways; alertness increases, the pupils widen in order to let in more light, the adrenal glands begin to pump more adrenaline and other hormones into the bloodstream, the heart races, the muscles tense, the blood pressure rises, digestion slows, the liver converts starches to sugar to

---

[8] If Madeleine studied meditation, however, she might be able to exert some control over these processes.

generate more energy, and sweat production increases, sometimes leaving the hair of our bodies standing on end. (2006, 310)

What conclusions follow from this phenomenological discussion of affect? Clearly, the analysis does not tell us anything in principle about whether computers will ever learn to talk about human emotions if they lack human-like bodies. But, this was not the point of the discussion. The analysis does show that there is at least one dimension of human experience—affect—in which Madeleine and the rest of us *de facto* experience sensations throughout our entire bodies, even though inputs from the brain are causally relevant.

Does this point have any purchase on the issue of linguistic socialization? Yes. If Madeleine can tell stories in which she displays an ability to create scenarios in which she empathizes with her characters, or if she provides an account of someone else's experience in an empathetic manner, it would seem that she is only capable of doing so because she had already developed a first-person understanding of what affect is—an understanding that came about through deeply embodied processes. Likewise, when Madeleine converses about experiences where an object takes on an emotional-symbolic significance, perhaps an instance where two people fight over who gets to sit at the head of the table, she also seems to be drawing from a frame of reference that she developed through first-person affective experience. This is not to say that Madeleine needs to have experienced a particular emotion in order to discuss it. If Madeleine has never been in love, then, as Collins would assert, she could still speak about it, and maybe even compose moving poetry on the subject. *But if Madeleine did not personally experience a range of emotions first-hand, it is hard to imagine how she could come to talk about a range of emotions in a skillful way that would pass a Turing Test.* In other words, Madeleine knows how important her wheelchair is to her, and she can *extrapolate* from that experience to talk about why other people would be emotionally invested in sitting at the head of the table. If Madeleine was never in love, she could still *draw from associated experiences*, pleasures, pains, longings, etc., that are common to her everyday openness towards the world. My main point, then, is that Collins fails to inquire into the complex developmental relations that obtain (in human experience) between embodied perception and higher-level cognitive abilities, including language development and use. I am suggesting—as does Dreyfus as well as the linguist George Lakoff and philosopher Mark Johnson (1999)—that Collins should even consider the possibility that "abstract concepts and abstract logic ultimately can be reduced to concrete, sensorimotor structures" (Brey 2001, 51). If Collins disagrees with this view, he owes us an alternative account of how Madeleine could make empathetic conversation in the absence of having empathetic experiences herself, or in the absence of having similar first-hand affective experiences that she could extrapolate from. In earlier essays, I even provided reason to believe that Madeleine's basic understanding of time and space—fundamental categories for *perceiving and talking about* matter and motion in our world—derive from her embodied orientation to movement and perception (Selinger 2003; Selinger-Mix 2006). Collins has yet to show why this analysis is wrong.

Finally, we should not lose sight of the fact that even though Madeleine is blind, she can construct a *body image*. Madeleine knows all-too-well that people can—and, perhaps, often do—respond to her in ways that place emphasis upon her bodily limitations. Considering how much assistance her aids provide, Madeleine's body image is probably rooted in the *cognitive and affective* significance that she accords to "dependency" and "independence". In developing her own body image, Madeleine also enhances her framework for understanding others. She becomes aware of how humans establish identity by judging other people's bodies—people who, in turn, evaluate our own corporeality. This understanding—one that became theoretically popularized in Jean-Paul Sartre's existential phenomenology—is an integral component in her ability to make sense of other people's actions—and, ultimately, to talk about other people's actions in an intelligible and empathetic manner.

Collins could object to my position by claiming that I conflate correlation with causation. He could reply that Madeleine's first-person embodied experiences of affect are causally irrelevant to her ability to make empathetic conversation. While longstanding patterns and stages of human development–cognitive and perceptual–leave me skeptical of this possibility, I concede that my view is falsifiable if the right kind of empirical tests prove otherwise. However, none of Collins's remarks about empathy or linguistic socialization are compelling in this respect; no alternate developmental evidence is provided. I would thus invite him to find a case in which a person was deprived of all affective experiences and still succeeded in becoming a conversationalist who could pass a Turing Test on human affairs. To crystallize the stakes of this invitation, I will return to the issue at the essay's conclusion.

Before proceeding to the colorblind experiment, it is important to note that Collins is not alone in displaying insufficient sensitivity to the complexity of whole-body perception. As it turns out, Sacks himself makes some of the same errors that Collins does when titling Madeleine's story "Hands".[9] What I want to question, then, is whether Sacks's literary inclinations and choices influence how he tells Madeleine's story. Because Sacks wrote the essays compiled in *The Man Who Mistook His Wife for a Hat* for a general audience, one needs to interpret his claims carefully. Firstly, the studies that appear there contain significantly less scientific detail than articles that appear in medical journals. Secondly, Sacks does not try to advance scientific knowledge in that book; his aim is to get a large audience to be engaged with abnormal human behavior.

Sacks is puzzled when he first examines the 60 year old Madeleine because she was unable to recognize objects placed in her hands even though the sensory capacities of her hands were "completely intact: she could immediately and correctly identify light touch, pain, temperature, passive movements of the fingers" (1998, 60). Sacks concludes that in order to get Madeleine to be capable of perceiving with her hands, he first needs to get her to discover her hands—that is, he needs to create the conditions under which Madeleine would initiate using her fingers to explore and perceive her environment. In other words, Sacks wants to find a way to get Madeleine to experience an "impulse" to use her hands interrogatively (1998, 61).

Sacks's solution is to change Madeleine's feeding conditions. He reasons that since Madeleine has been taken care of her entire life, she has not been given an incentive to use her hands. After instructing the nurses to move Madeleine's food "slightly out of reach on occasion," a hungry Madeleine commits "her first manual act" by reaching out and grabbing a bagel (1998, 60). Sacks characterizes this event as one that marks Madeleine's "birth as a complete 'perceptual individual'" (1998, 61). Because Madeleine

---

[9] I am grateful to John Mix for taking the time to revisit Sacks's assumptions in the Madeleine case. The analysis of her situation and my ability to grasp the austerity of Collins's minimal embodiment thesis would not be possible without his input.

actively grabs something with her hands for the first time, Sacks emphasizes how she comes to recognize objects:

> A bagel was recognized as round bread, with a hole in it; a fork as an elongated flat object with several sharp tines. But then this preliminary analysis gave way to an immediate intuition, and objects were instantly recognized as themselves, as immediately familiar in character and 'physiognomy,' were immediately recognized as unique, as 'old friends.' And this sort of recognition, not analytic but synthetic and immediate, went with vivid delight, and a sense that she was discovering a world full of enchantment and beauty. (1998, 61)

Much of Sacks's description here accords with a phenomenological orientation. Like phenomenologists, Sacks describes human perception as a skill that is developed through motivated action. Like phenomenologists, Sacks links the human impulse to develop skill as laden with affect, i.e., Madeleine's impulse is initiated by frustration and it continues by being fueled with delight. Finally, like phenomenologists, Sacks describes Madeleine's perception of objects as an intuitive, synthetic, and immediate act. Dreyfus constantly makes these same points, and like Sacks he frames these observations as contrasting with "analytic" conceptions of perception.

What, then, do I find objectionable in Sack's account? Sacks seems to suggest that the reason why Madeleine could recognize a bagel as a bagel is because she had already developed an understanding of what a bagel is based upon stories that were told to her. Sacks writes: "Had she not been of exceptional intelligence and literacy, with an imagination filled and sustained so to speak, by the images of others, images conveyed by language, by the *word*, she might have remained almost as helpless as a baby" (1998, 62). In contrast to the linguistic images that Madeleine had obtained, Sacks notes that she lacked the simplest internal tactile images to draw from (1998, 62). The implication of Sacks's discussion is that Madeleine first acquired an understanding of bagels from conversations about them, and then—once she finally touched a bagel with her hands—extrapolated from verbal understanding to tactile understanding. In other words, for Sacks the medium of conversation had already given Madeleine a conceptual understanding of "bagelhood"; all she needed to do was apply this understanding in a tactile manner by correctly identifying a concrete object as a bagel.

But what justifies this conclusion? To be sure, Madeleine's linguistic instruction certainly played a significant role in her ability to develop a new skill. I agree with Sacks's view that without already possessing images conveyed by words, Madeleine would, in all likelihood, have been unable to develop the requisite perceptual recognition via her hands as quickly as she did. But why assume that because Madeleine could not initially recognize objects by touching them that she, therefore, failed to compensate, as best she could, for her diminished perceptual capacity by taking advantage of as much perceptual information as she could avail herself of? By Sacks's own description, the sensory capacities of Madeleine's hands were "completely intact." Here, we should recall that the experience of touch is neither localized to the hands nor to the brain; it is a human experience that is distributed throughout her entire body. For all human beings, touch occurs wherever the skin's nerve endings transmit sensations to the brain, and it can even be affected by the presence of hair. Moreover, human skin turns out to be much more biologically complex than anything stipulated in the minimal embodiment thesis. According to the Texas Education Agency (April 2004), each square inch of skin contains approximately:

78 nerves,
650 sweat glands,
19-20 blood vessels,
78 sensory apparatuses to detect heat,
13 sensory apparatuses to detect cold,
1,300 nerve endings to record pain,
19,500 sensory cells at the end of nerve fibers,
160-165 pressure apparatuses for the sense of touch,
95-100 sebaceous glands,
65 hairs and hair muscles,
19, 500,000 cells.

Given these considerations, while I can agree with Sacks that prior to developing the motivating impulse, Madeleine could not identify a bagel if one were put in her hand, I take issue with the (assumed converse) idea that her only understanding of what a bagel is came from language. If Madeleine had eaten a bagel before, she clearly took perceptual stock of it through her mouth, gums, tongue and lips; the resistance it offered to her teeth was significant, as was the sound it made while being chewed and the sensations it provided her nose while being smelled.[10] From a developmental perspective, it is important to note a few things. Hearing is a more mature sense at birth than vision, and oral and olfactory exploration play a crucial role for babies; while the tongue and gums are vital for exploring surroundings, it may even be the case that newborns can recognize their mothers by scent. Our lips, moreover, are extraordinarily receptive surfaces. Some people have even theorized that humans often express intimacy by kissing each other on the lips because of how sensitive the surface is. Thus, while Madeleine may have had, comparatively speaking, a diminished perceptual sense of what a bagel is because she could not see one through her eyes or take in sophisticated information about it through her hands, she nevertheless could still develop rich sensory awareness of what a bagel is through the incarnate processes of exploring and eating them. And, for all we know, it may even be a mistake to characterize Madeleine's sensory awareness as diminished. Certainly, her awareness is *different* from the experiences of able-bodied people. However, neither Collins nor Sacks informs us as to whether Madeleine has developed heightened perceptual sensitivities as compensation—or, a kind of "sensory substitution"—for her physical impairments. Alva Noë and some colleagues even suggest that auditory awareness can be experienced qualitatively as visual perception: "For example, a woman wearing a visual-to-auditory substitution device will explicitly describe herself as seeing through it" (Humphrey 2006, 57).

In addition to the perceptual information that Madeleine could take in about objects in general, we need to remember that with respect to the example under review she could *tacitly compare* the experience of eating a bagel to eating other foods. And, if she was given a bagel that was sliced into digestible pieces, she could ask whoever was feeding her to describe what the bagel was like before it was cut up. In this context, her understanding of what a bagel is seems to develop through *discursive and embodied* means. Moreover, we need to remember that the first two objects that Madeleine identifies, a bagel and a fork, are objects that are presented to her in a *specific context*: Madeleine knows that the objects in front of her are objects that she would encounter at mealtime. Meals are significant to Madeleine because she is

---

[10] We are given no reason to believe that Madeleine had not already eaten many bagels in her 60 years of life.

aware of how uncomfortable it is to be hungry, and the relevant class of objects that she is presented with during mealtime is quite limited in comparison to the entire class of objects that exist. By not reflecting on the context in which Madeleine first recognizes an object after touching it, Sacks conveys the misleading impression that Madeleine is doing something akin to playing a children's game where completely random objects are placed in a box and kids compete over who can identify them through touch.

Since Sacks does not describe the contexts in which Madeleine comes to identify other objects correctly, and since he does not discuss the relevance of Madeleine's ability to take in diminished information about objects through her intact and compromised senses and skin, he fails to establish that her perceptual inferences are rooted exclusively in extrapolations from internal images that were formed through conversation. Were Sacks to have provided a more nuanced view of how Madeleine's perceptual awareness is distributed throughout her body, it would have been more diffi-cult to endorse the dramatic chapter title, "Hands." In the spirit of Collins's minimal embodiment thesis, in this one story Sacks conveys the misleading impression that the cognitively relevant human sense of touch can be localized to one body part, the hands.

4b Colorblindness and Gravitational Wave Physics Revisited

In the last section, I emphasized that by focusing on Madeleine as an adult, Collins and Sacks both failed to inquire into the relevance of her embodiment for her development as a skillful speaker. My main point about colorblindness is merely an elaboration of this phenomenological trajectory.

When Collins describes his colorblind experiment, he notes that the colorblind can learn to talk about colors they never perceive directly. While I am not skeptical about this display of skill, I take issue with Collins's view that the experimental data estab-lishes that bodily ability is not a necessary condition for linguistic ability. Collins draws the wrong conclusions not because he cal-ibrates his controlled experiment in terms of four test subjects, i.e, colorblind, color-perceiving, pitch-blind, and pitch-perceiving subjects. Given his assumptions, this configuration is quite in-genious. Collins errs for other reasons—reasons that he himself cites in the Appendix.

In the Second Appendix to the colorblind paper, Collins writes:

Finally, colorblindness and perfect pitch have been
discussed as though they were binary qualities—one
either has them or one does not. But they each come
in different types and each type lies on a continuum.
To get a strong effect in the experiment participants
need to be located toward the extreme end of the spectrum.

Unfortunately, this point is too important to list in an appendix
as merely a "confounding factor". As Collins notes: "The answers
given by such a minimally colorblind person are likely to be indistinguishable from that of a color-perceiver under any test; likewise,
a judge who is only marginally colorblind in this way does not
really possess colorblindness as a 'target expertise.'" Based upon
these considerations, Collins errs in conceiving of the current colorblind study as research that provides data which validates the
minimal embodiment thesis. *As currently configured, the test does
not tell us anything about the analogical and inferential capacities of the participants.* No measures are taken to determine how
much of the participants' ability to talk about color can be linked
to their ability to extrapolate from the colors that they are capable of perceiving. Moreover, no measures are taken to determine
how much of the participants' ability to talk about color can be
linked to their ability to extrapolate from conceptual frames of reference that were acquired through embodied forms of interaction
and perception during early developmental stages.

Furthermore, even if the participants could only see in black-
and-white, Collins would still need to investigate whether their
ability to talk about colors arose developmentally through physical activity and distributed perceptual experience. Were he to do
so, I bet he would discover compelling evidence suggesting that
sensorimotor skills and distributed perception proved crucial to
their ability to form the concept of color. It would not be surprising if it turned out to be the case that before learning what
color is, babies use emotionally motivated physical exploration
and sensory information to learn how to distinguish themselves
from objects. By the time that babies are taught that color is an
attribute, they already have learned to use physical exploration
and sensory information to develop a rich understanding of their
world. Even a completely blind person like Madeleine can learn
crucial lessons about what color is by being instructed through
contrasting perceptual exercises, i.e., by being taught that while
color is not something one tastes, smells, feels, it nevertheless

resembles these qualities in that it is a perceptually descriptive aspect of phenomena. In this context, we should add that while the vOICe technology mentioned in the previous section might be capable of conveying what colors it detects, it cannot teach anyone what color is.

Ultimately, there are many perceptual skills that humans develop when learning to talk about primary colors. We have to learn to ignore relative motion, learn that there is a sense in which the color of an object remains constant even when modifications are made to that object (e.g., a red apple remains red even as it transitions form being uneaten to being partially eaten), learn that there is a sense in which color is separate from other characteristics (i.e., size, function, etc.), learn that color has a normative dimension (i.e., learn that since lighting can influence how color is perceived, it is sometimes necessary to move around until ideal lighting conditions obtain), learn that imaging technologies can change the appearance of color, etc. Again, I would contend that much—if not all—of this learning typically occurs *tacitly through physical interactions with objects and through perceptual experiences that are distributed throughout the socialized body.* If I am wrong on this point, Collins should provide an alternative developmental account. To assist Collins in this process, I invite him to answer the following questions:

1. What activities did Madeleine likely engage in when she first learned how to recognize herself as a being who is separate from the objects she encounters?

2. Did any of these activities involve physical interaction with herself, other people, or objects? If so, what kinds of physical interaction took place? And, what it was it like for Madeleine to experience such physical interactions? Were all of her experiences localized in her brain, or were some felt throughout her body (i.e., through a complex feedback system that comprises the nervous system, senses, skin, glands, muscles, etc.)?

3. How might learning to recognize herself as separate from the objects she encounters be a useful prerequisite for Madeleine learning to talk about color?

4. What activities did Madeleine likely engage in when she learned to talk about color?

5. Did any of these activities involve physical interaction with herself, other people, or objects? If so, what kinds of physical interaction took place? And, what it was it like for Madeleine to experience such physical interactions? Were all of her experiences localized in her brain, or were some felt throughout her body (i.e., through a complex feedback system that comprises the nervous system, senses, skin, glands, muscles, etc.)?

6. Does the narrative provided when answering questions 1-5 fit within the parameters of the minimal embodiment thesis?

If Collins provides plausible replies, he should answer the last question in the negative.

Having said this, it should be clear that I am not making any in-principle claims about whether disembodied machines will ever learn to talk about primary colors in the way humans can. *All that I am claiming is that the de facto practices humans participate in when they learn to talk about color are practices in which discursive training, embodied exploration, and embodied experience all prove to be crucial components.* This modest claim, however, is sufficient to refute Collins interpretation that the colorblind experiment disproves phenomenological claims about how humans use their bodies to become socialized into color discourse. Collins provides no reason to believe that any of the participants in the colorblind experiment learned to talk about color by participating in processes that differ from the ones I have discussed.

By refuting Collins's interpretation of embodiment in this experiment, I diminish substantively the amount of available empirical evidence that he can appeal to in order to validate his minimal embodiment thesis. If my claims in the last section were sufficient, then Collins is not entitled to appeal either to the Madeleine case or the colorblind experiment. His only remaining evidence is his analysis of his success at mastering the discourse of gravitational wave physics. But based on all the analysis that has transpired thus far, it is not necessary to discuss that case in detail. From a phenomenological point of view, we need to remember that Collins is an able-bodied sociologist who masters the discourse of gravitational wave physics. While it is remarkable that he can do so in the absence of *experience performing gravitational wave experiments*, Collins never addresses the basic developmental question: If he did not already have a basic understanding of the world that he developed through *linguistic and embodied means*, would he have developed an appropriate background understanding that would enable discussions about gravitational wave physics to be perceived as meaningful?

Put otherwise, comparing himself with Madeleine and colorblind people is an ill-conceived program, precisely because the comparison is predicated upon a misunderstanding of the conditions of embodiment that prevail. While Collins may lack direct experience performing gravitational wave physics experiments, he is not in any significant way physically challenged. By contrast, Madeleine and colorblind people lack physical capacities that able-bodied people possess. To inquire into the conversational capacities of Madeleine or colorblind people is thus to inquire into how they managed to overcome physical disabilities that limited how they interacted perceptually with the world. But to inquire into Collins's own conversational capacities is to inquire into how he managed to overcome the limits of his disciplinary training. In other words, Collins never asks: How, as an able-bodied person with an expert's knowledge in the sociology of science and technology, was he able to extrapolate from what he knew and what he experienced to develop expert ability at talking about gravitational wave physics? To begin to answer this question, Collins would need to provide a phenomenological account of his interactions with gravitational wave physicists and gravitational wave physics books. It would be surprising if the relevant training were entirely discursive—that is, if body language, communicative gestures, pictures of scientific equipment, and maybe even direct perceptions of laboratory equipment were anything less than critical. Moreover, Collins would also need to consider the impact of his background common sense (acquired, in part, through embodied learning) and his background technical knowledge of science. Here, explicit consideration of Collins's inferential and analogical capacities, and his development of those capacities, would need to be provided. In effect, Collins underestimates the richness of what he himself already brings to the conversational table.

Conclusion

Do Collins's experiments teach us anything? Yes. Collins provides compelling evidence that contributes directly to debates about whether human beings can pass Turing Tests on topics that they have not directly experienced. This contribution is relevant to both philosophical and sociological attempts to understand expertise. As Collins realizes, it may enable humanists, social scientists, and natural scientists to forge new directions past the academic "science wars."

Unfortunately, Collins wants to make a stronger claim than the one just reconstructed. Collins insists that he establishes something important about embodiment. He claims that given how much information the medium of conversation can convey, a minimally embodied agent with sufficient conversational experience can pass a Turing Test on subjects that the agent has not experienced first-hand.

The cases that Collins studies fail to support this conclusion. As a result of having developed to cognitive maturity, all of his test-subjects combined physical interaction and neural-somatic perception to develop frameworks that they later tacitly appealed to when learning to talk about topics they have not experienced. Collins never establishes, and indeed never tries to establish, that his test subjects could still be masterful speakers if they lacked any rich, antecedent corporeal apprenticeship. Indeed, based on the data that Collins selects for study, he cannot support this conclusion; only adult test-subjects are examined, and these subjects have already undergone corporeal apprenticeship before Collins examined their linguistic skill. Ironically, then, Collins claims to be providing an account of what can be learned through conversation, but he never examines how human beings actually learn to converse. In bypassing the phenomenological issue of skill development, Collins theoretically invents a developmentally unsound view of conversational practice. It is as if he reasons that if someone can talk about a topic that they have never experienced, the only way that the person could compensate for their experiential deficiency is to utilize the fullest potential of discourse alone. On the contrary, I hope to have shown Collins presents insufficient evidence for establishing this position. In the cases considered, conversation functioned as an important part of the compensatory process, but it depended upon the reception of a non-minimally embodied agent.

In short, what Collins calls "linguistic socialization" is really a developmental process in which agents avail themselves of discursive instruction, physical interaction, and acts of perception that are not localized in the brain. Given this range of components, Collins should drop the term. It misleads by suggesting that conversation itself can socialize (in the Turing Test sense) an unsocialized person. Of course, it would be acceptable for him to replace the phrase with "embodied-linguistic socialization." But this cumbersome locution is unnecessary. It is unnecessary because the thrust of my analysis suggests that Collins should describe his research as inquiry into what can be said in the absence of "direct experience"; his project to undermine phenomenological embodiment should be abandoned.

Should Collins want to proceed further with his research program, then he needs to be clearer about how to construct an appropriate empirical test for his idea of minimal embodiment. Unfortunately, the ideal test would be so coercive and vile that it should never transpire. Collins would have to take a human baby straight from birth and immediately shut down that baby's consciousness by inducing a comma (or comma-like state). While in that comma, the baby would need to be transplanted into a sensory-deprivation machine that could constrain the baby's perceptual ability to the conditions specified by the minimal embodiment thesis. Thus, if the baby had the experience of floating in water, the machine would be insufficiently restrictive. If under these austere conditions Collins could speak to the baby, and over time use discourse to get the baby to be a skillful conversationalist, then, and only then, would he be capable of proving his minimal embodiment thesis. Are there any alternatives? None that I can think of. While feral children are removed from society, they still have fully functioning human bodies with which to explore their surroundings.

As a final note, I would like to point out that Collins and the GOFAI researchers have more in common than Collins recognizes. Collins thinks that because he and Dreyfus agree that tacit knowledge cannot be formalized according to GOFAI parameters, he and Dreyfus both agree on why conversations are skillful events. What Collins does not appreciate is that Dreyfus is not interested solely in the tacit dimensions of skillful actions. For Dreyfus, in order to understand the tacit dimensions of human skill, one needs to understand the tacit dimensions of embodied action and embodied perception that allow humans to develop their skills in the first place. In this respect,it can be said that both GOFAI researchers and Collins take an overly-intellectual, abstracted, and informational approach to the phenomena of language. Had Collins used resources compatible with phenomenology to inquire into why it is that humans can learn as much as they can from conversation, he would have recognized that the minimal embodiment thesis obscures, rather than points the way towards the answer. Instead, just as the computer science background of the early GOFAI researchers inclines them to *imagine* that humans are, at bottom, a bunch of rules and programs, Collins's sociological training seems to incline him to *imagine* that crucial forms of human learning are, at bottom, mostly derived from conversation.

Bibliography

Brey, Philip. 2001. "Hubert Dreyfus: Humans versus Computers." In *American Philosophers of Technology: The Empirical Turn.* Ed. Hans Achterhuis. Bloomington: Indiana University Press.

Collins, Harry. 1990. *Artificial Experts: Social Knowledge and Intelligent Machines.* Cambridge: MIT Press.

————. 1992. "Dreyfus, Forms of Life, and a Simple Test for Machine Intelligence." *Social Studies of Science* 22: 726–39.

————. 1996. "Embedded or Embodied? A Review of Hubert Dreyfus' *What Computers Still Do.*" *Artificial Intelligence* 80: 99–117.

————. 2000. "Four Kinds of Knowledge, Maybe Three Kinds of Embodiment, and Question of Artificial Intelligence." In *Heidegger, Coping, and Cognitive Science: Essays in Hubert L. Dreyfus,* Vol.2. Ed. Mark Wrathall Jeff Malpas. Cambridge: MIT Press.

————. 2001. "Tacit Knowledge, Trust, and the Q of Sapphire." *Social Studies of Science* 31 (1): 71–85.

————. 2004a. "interactional expertise as a Third Kind of Knowledge." *Phenomenology and the Cognitive Sciences,* 3 (2): 125-143.

————. 2004b. "The Trouble with Madeleine." *Phenomenology and the Cognitive Sciences,* 3 (2): 165-70.Collins, Harry and Robert Evans. 2002. "The Third Wave of Science Studies: Studies of Expertise and Experience." *Social Studies of Science,* 32, (2): 235-296.

Collins, Harry and Robert Evans. 2007. *Rethinking Expertise,* Chicago: University of Chicago Press.

Collins, Harry, Robert Evans, Rodrigo Ribeiro, and Martin Hall. 2006. "Experiments with Interactional Expertise." *Studies in History and Philosophy of Science* 37a (4).

Collins, Harry and Martin Kusch. 1999. *The Shape of Actions: What Humans and Machines Can Do.* Cambridge: MIT Press.

Dusek, Val. 2006. *Philosophy of Technology: An Introduction.* Massachusetts: Blackwell Publishing.

Dreyfus, Hubert. 1965. "Why Computers Must Have Bodies in Order to Be Intelligent." *Review of Metaphysics* 21:13-32.

———. 1992a. *What Computers Still Can't Do*. Massachusetts: MIT Press.

———. 1992b. "Response to Collins, *Artificial Experts*." *Social Studies of Science* 22: 717–26.

———. 2000. "Response to Collins." In *Heidegger, Coping, and Cognitive Science: Essays in Honor of Hubert L. Dreyfus*, vol. 2. Ed. Mark Wrathall and Jeff Malpas. Cambridge: MIT Press.

———. 2001. *On the Internet*. New York: Routledge.

Dreyfus, Hubert and Stuart Dreyfus. (1986). *Mind Over Machine: The Power of Human Intuition and Expertise in the Era of the Computer*. New York: Free Press.

Humphrey, Nicolas. 2006. *Seeing Red*. Cambridge: Belknap Press of Harvard University Pres.

Lakoff, George, and Mark Johnson. 1999. *Philosophy in the Flesh: The Embodied Mind and Its Challenge to Western Thought*. New York: Basic Books.

Sacks, Oliver. 1998. *The Man Who Mistook His Wife for a Hat and Other Clinical Tales*. New York: Simon and Schuster.

Sanders, John T. 1985. "Experience, Memory, and Intelligence." *The Monist* 68 (4): 507-521.

Selinger, Evan. 2003. "The Necessity of Embodiment: The Dreyfus-Collins Debate." *Philosophy Today* 47 (3): 266-279.

Selinger, Evan and Robert Crease. 2002. "Dreyfus on Expertise: The Limits of Phenomenological Analysis." *Continental Philosophy Review* 35: 245-279.

Selinger, Evan and John Mix. 2004. "On Interactional Expertise: Pragmatic and Ontological Considerations." *Phenomenology and the Cognitive Sciences* 3 (2): 145-163.

Sheets-Johnstone, Maxine. 1986. "Existential Fit and Evolutionary Continuities." *Synthese* 66 (2): 219–48.

———. 1990. *The Roots of Thinking*. Philadelphia: Temple University Press.

———. 1999. *The Primacy of Movement*. Amsterdam: John Benjamins.

Thompson, Evan. 2001. "Empathy and Consciousness." *Journal of Consciousness Studies* 8 (5- 7): 1-32.

Todes, Samuel. 2001. *Body and World*. Cambridge: MIT Press.

4

The Incompatibility of Industrial Age Expertise and Sustainability Science

Written with Thomas Seager

Introduction

The concept of "wicked problems" (Rittel & Webber 1973) was one of many powerful new ideas that thematized connections between the environment, society, and economy in the public and expert consciousness during the late 1960's and early 1970's when the post-War generation began to mature. Until then, the dominant technological themes that drove the Industrial Revolution were the specialization of labor (including intellectual labor) and the reduction of complex systems to simplified principles that enabled isolated experimentation. In response to the perceived failure of industrial-age science to solve such problems as poverty, crime, and public education, Horst Rittel and Melvin Webber (1973) joined other leading authors of the time by identifying interconnectivity and complexity as decisive features of social planning problems which pose a central challenge to science itself. In general, they advocated a new approach that placed greater emphasis on deliberation over operations analysis, participation with multiple stakeholder groups, social learning, and adaptation accompanied by skepticism regarding the limits of science and technology. In retrospect, it seems inevitable that the leading concepts emerging during this time came to be recognized as themselves interconnected. Now they are subsumed under one of the most pervasive buzzwords of the early 21st century: *sustainability.*

The most common understanding of sustainability derives from the Bruntland Commission (Bruntland & Khalid 1987) definition of *sustainable development*, which can be paraphrased as development that meets the needs of present generations without impoverishing future generations. In this context, sustainability is interpreted in narrow economic terms, as a problem of natural resource stewardship. However, the economic perspective cannot be absolute. Contexts exist in which sustainability requires understanding of complex global systems in which cause and effect relationships are extraordinarily difficult to decode, deliberation is encumbered by contrasting value judgments or cultural norms that are far from universal, and the relevant knowledge resides in multiple disciplines that lack streamlined techniques of collaboration. Moreover, because the ability of future generations to meet their own needs partly depends upon technological innovation, debates over how to operationalize sustainable parameters remain tethered to divisive views regarding the relations between science, technology, and progress. For these and related reasons, sustainability is best understood as an *essentially contested concept* (Connelly 2007).

A fundamental conceptual connection between sustainability and wicked problems is attributable to Bryan Norton (2005), who argues that sustainability problems typically exhibit the ten characteristics (which, in principle, can be reduced to five) that are constitutive of wicked problems: difficulties in problem formulation, multiple but incompatible solutions, open-ended timeframes, novelty (or uniqueness), and competing value systems or objectives. Despite Norton's compelling characterization, *scientific organizations rarely characterize sustainability in "wicked" terms*. While the Bruntland Commission report explicitly incorporates environmental, economic, and social concerns that are clearly connected to the problems of planning that Rittel and Webber's original theory focused on, the logic of wicked problems moves expertise beyond the scope of the Bruntland Commission *because it calls to attention to the limits of industrial-age science*. In other words, the Bruntland Commission, along with most university programs in sustainable science and engineering, merely gesture to sustainability being a post-industrial concept. In practice, their outlook expresses confidence in the industrial science paradigm.

This gap between knowledge and practice is partly a residue of the Industrial Age itself. Prior to the publication of Rittel and Weber's seminal work on wicked problems, Martin Heidegger (German original1954, English reprint 1977) wrote "The Question Concerning Technology" in which he famously argued that modern science is guided by an engineering ethos driven by the values of manufacturing and production. The basic idea is that modern science co-evolved in close concert with industry. As a consequence, during the last two hundred years scientific emphasis has centered on *technology* –or, when understood literally, the study

of *technique*. This emphasis has led to an ossified public imagination that equates progress with making things cheaper, smaller,
faster and, in general more efficient. Rittel and Webber themselves point to water treatment, transportation, education, and
health care infrastructure as exemplary cases of scientific achieve-
ment. Following the line of thought opened up by Heidegger, we
could just as easily draw upon more modern examples from personal electronics, telecommunications, and computing. While the
advances in these areas of science and technology during the last
forty years have been extraordinary, it is not yet clear that they
have contributed significantly to problems of social progress. At
the core, the concept of wicked problems points out that a theory
of planning "ain't rocket science"—in fact, it's much harder. The
obvious implication is that a *new type of scientific expertise is required for a "post-industrial" age* (e.g., Bell 1973). Nonetheless,
the scientific community, despite being quick to adopt the rhetoric
of sustainability as a legitimate goal of science, has been slow to
accept the critique of science on which a deeper understanding of
sustainability reposes.

1. A Taxonomy of Sustainability Science

To clarify fundamental differences between understandings of sustainability that confront and ignore wicked problems, we offer here
an idealized taxonomy that identifies three general approaches to
sustainability: "business as usual," "systems engineering," and
"sustainability science" (Table 1). Although simplifying strategies
are needed to construct this taxonomy, it accurately captures our
main point: *only sustainability science, which is in the minority,
is informed by an understanding of wicked problems.*

Business as Usual

The most common approach to sustainability is to simply repack-
age the normal incremental practice of research and development
as pertaining to making the world in general a better place –
and, as a consequence, relevant to sustainability. This approach
is typically extraordinarily optimistic about the prospects of technology to improve the human condition. That is, it presupposes
that efficiency increases or the introduction of new capabilities will
necessarily improve environmental, social, and economic quality.
However, the business as usual approach typically ignores environmental and social issues altogether, and may even ignore relevant
economic considerations. Ultimately, the emphasis in business as

Table 1: Science & Technology Orientations Towards Sustainability

	Attitude towards Technology	Focus	Expert & Ethical Culture	Approach to Complexity	Approach to Conflicting Views
BUSINESS AS USUAL	Optimism	Creating new things, resources. Ignores scale & efficiency.	Depth in a single sub-discipline. Professional ethics.	Simplification, reduction	Defense of techno-industrial ethos. Denial of opposing perspectives.
SYSTEMS ENGINEERING — Engineering within ecological constraints	Pragmatism	Cost optimization of maturing technology. Ignores scale.	Compartmentalized, multi-disciplinary teams. Social ethics.	Cost-benefit optimization.	Litigation, regulation.
Sustainable engineering		Optimization for triple bottom line. Ignores scale.		Risk minimization	Structured participation.
SUSTAINABILITY SCIENCE	Skepticism	Sustainability as a wicked problem	Transdisciplinary interaction. Macro ethics.	Adaptation, resilience.	Cooperation, deliberation.

FIGURE 4.1.

usual is on creating knowledge that leads to new capabilities, regardless of broader contextual questions.

For example, the discovery of carbon nanotubes (CNT) by Sumio Iijima (1991) set in motion a multi-billion dollar research and development effort directed towards characterization and engineering of CNT for applications in wiring, catalysis, structures, and electrochemistry. Far less effort has been expended in the area sustainability concerns, such as health and life-cycle environmental effects – despite the fact that some CNT are currently more energy-intensive in manufacture than even the most high-tech and demanding materials, including crystalline silicon semiconductors (Healy et al. 2008), and may be hazardous under certain exposure scenarios (Oberdörster et al. 2007, 2007).

While the business as usual approach may lead to scientific breakthroughs in materials, medicine, energy systems and other technologies that generally correlate with human progress, objecters claim it yields advances in technology that disproportionately benefit the rich and thereby exacerbate social inequality. The usual response to this critique is two-pronged:

1. Since the scientific enterprise is capable of generating technologies appropriate for underprivileged classes or underdeveloped countries, the fault lies not with science and technology *per se*, but their application; and,

2. Since the poorest people in industrialized countries today are far better off than typical, pre-industrial populations, the *disparity* between rich and poor is not as important as the *absolute level* of well-being among the poorest.

A minority—but certainly not inconsequential—view, espoused by Heidegger-inspired critics like Albert Borgmann (1984), is that the business as usual approach is morally problematic when it is driven by and reinforces consumerism. Borgmann characterizes consumerism as a spectator's orientation to life that is fostered by the dominant modern technological trajectory called the "device paradigm"—a paradigm that putatively seduces people to passivity by addicting them to the artifacts of technology that separate means from ends and strip away meaning. Such addiction, Borgmann insists, undermines "focal" activities that are anchored in context and tradition, and which require skillful activity to yield memorable experiences where psychological "flow" abounds. As Aidan Davison (2001) notes, Borgmann's perspective suggests the business as usual approach can accommodate some sustainable ends even while remaining divorced from the virtue ethics goal of creating cultures that are guided by practices that provide genuine psycho-social-spiritual *sustenance.*

Systems Engineering

A more modern approach to sustainability involves what may generally be described as systems engineering. Two perspectives dominate this domain, both of which seek improvement at the scale of integrated systems, rather than piecemeal component optimization. The first outlook is *engineering within sustainability constraints.* In this approach, engineering systems are typically optimized for traditional objectives, such as cost minimization or rate of return maximization, but under more highly constrained conditions than have historically been the case. Environmental emissions standards – both regulatory compliance and voluntary standards that exceed compliance (such as L.E.E.D. certification for green buildings) – provide illustrative examples, as does the increased interest in stakeholder and public participation at early stages of engineering design development.

The second perspective goes even further than the first by expanding the design objectives themselves to incorporate the *triple bottom line* of sustainability: economy, environment, and society. In this approach, environmental quality and social objectives are not merely constraints to be met. They are understood as design objectives in their own right that necessitate assessment of trade-offs with respect to one another and cost. For example, the savings in fuel costs that result from hybrid automobiles may not justify the increased purchase price of the technology under all but exorbitant fuel price scenarios. Nevertheless, hybrid autos provide environmental benefits in the form of reduced tailpipe emissions – especially in congested urban areas that are most impacted by poor air quality. These benefits partially justify government programs that subsidize the private purchase of hybrid cars through tax credits (Keefe et al. 2008). More speculatively, the claim can be made that hybrid cars provide social benefits to the owners that perceive enhanced social standing in the community or derive satisfaction from a conspicuous display of environmental awareness, and that these social dimensions justify the increased expense. After all, it has long been a tradition in the auto industry to advertise cars based upon the importance drivers place on self-image. From a systems perspective, then, hybrid automobiles appear to exemplify technology that incorporates economic, environmental, and social considerations in design. Nevertheless, systems optimization approaches very rarely incorporate considerations of *scale*, and are therefore vulnerable to the criticism that the new technologies that make goods more efficiently (and therefore cheaper) will result in unsustainable growth in consumption. In several cases, historical examples support this critique (e.g., *The Coal Question*, Jevons 1865) by showing that increases in consumption occur simultaneously with increases in efficiency. The typical rebuttal is that technological substitution has driven downwards the price of almost all basic commodities and manufactured goods throughout the Industrial Revolution (with the possible exception of lumber – Simon 1996). To the extent that price is an indicator of scarcity, long-term declining prices would seem indicative of *increasing* abundance, despite concurrent increases in consumption.

Sustainability Science

At the extreme, sustainability science represents a paradigm shift in the way that science approaches problems of technology and complex systems. The term has previously been employed to describe science working at the boundaries of "industry and nature" (Clark and Dickson 2003), and to differentiate science that is "defined by the problems it addresses rather than by the disciplines it employs" (Clark 2007). We expand upon this understanding here by further describing the differences between sustainability science and other approaches to science. In a traditional, pre-industrial (or "normal") science approach, problems are defined narrowly and potential solutions circumscribed by that narrow definition. Assessing the relative merit of any particular technology is typically a matter of defining measureable performance objectives (e.g., dollar per watt installed capacity of photovoltaic systems, or central processing unit computational cycles per second) and defining success as achieving correlative policy or technology objectives (e.g., meeting Corporate Average Fuel Economy standards). The systems view described earlier introduces broader aspects of the problem such as may be suggested by a life-cycle perspective. However, in the domain of wicked problems, these approaches encounter several difficulties.

With regard to problem formulation, the evolving and recursive nature of wicked problems demands a constant adaptation in the goals of technology development as new information about feedback effects and unintended consequences is discovered. While it can be said that science in a business as usual paradigm is responsive to these discoveries, the ethical orientation, disciplinary approach and narrow focus of industrial-age science introduces obstacles and delays in that feedback that can exacerbate the unintended consequences of technological progress. For example, physicists and materials scientists developing CNT-based technologies typically lack the cross-disciplinary expertise necessary to understand toxicological and life-cycle environmental concerns. Similarly, toxicologists typically lack the specialized knowledge of nanomaterials required to fully characterize those properties of CNT that are germane to biological health responses or environmental fate. When considering sustainability concerns, the business as usual approach is significantly handicapped by the degree to which knowledge resides in increasingly narrow specializations. Ultimately, technological progress under the business as usual paradigm could *exacerbate* wicked problems, rather than contribute to their resolution.

Systems engineering represents an improvement on business as usual in that it explicitly attempts to incorporate broader, contextual concerns. Nevertheless, it remains deficient in several respects that result primarily from the focus of systems engineering on efficiency. The natural maturation of any new technology typically involves an incremental evolution from focus on new techniques in the business as usual approach, to optimization of manufacturing or life-cycle considerations. Optimization necessarily requires an objective function that provides the basis for comparing the overall merit of different design alternatives. Selection of any one objective criterion necessarily excludes others. Consequently, an optimization approach requires advancing one normative view of technology at the expense of others. As any particular engineered systems expands in terms of *scale*, it must increasingly encounter constraints or other interactions with both other engineered systems that have been advanced under competing ideals, and complex, adaptive natural systems. Two points of irreconcilable conflict arise.

The first is the conflict between different idealized visions of the engineered system, such as might be encountered in the context of climate change. For example, legitimate, value-laden disagreements with regard to the optimal levels of carbon dioxide in the atmosphere inevitably lead to different visions of the optimal technology platforms on which energy systems should be based. Conflicts between different views can not be reconciled on the basis of technical performance standards alone, which relates directly to the characteristic of wicked problems that can be described

as multiple or competing value systems, i.e., it is mathematic-
ally impossible to optimize a system for more than one objective
function.

The second point relates to complexity. Even if universal agreement could be attained on what constitutes the overall merit
of any technological alternative, the resulting engineered system
would (at scale) inevitably be subject to interactions with complex, adaptive natural and social systems. Emergent behaviors
and properties of complex systems mean that any optimization of
existing engineered systems is at best myopic (and at worst, ex-
acerbating risk of catastrophic collapse). That is, conditions only
appear optimal from the narrow perspective of the existing time-
frame. For example, the widespread adoption of transgenic crops
resistant to the herbicide glyphosate during the 1990's resulted in
economic benefits in terms of increased yields, as well as environmental benefits resulting from lower herbicide volumes and till-
age requirements. However, the recent emergence of glyphosate-
resistant weed species in many regions of the United States has
required a return to more traditional practices, such as crop-
rotation, tilling, and intensive application of aggressive herbicides.
In retrospect, the benefits of transgenic crop technologies could
have been extended if they had been deployed at a more limited,
albeit suboptimal, scale.

Sustainability Science does not necessarily reject business as
usual or systems engineering approaches as appropriate to problems that can be tamed. In this respect, sustainability science is
not fundamentalist. Its practitioners should not conceive of themselves as advancing research in something like the one true form of
authentic sustainability. However, sustainability science is guided
by recognition of the main reasons why the two main alternatives fail when confronting wicked problems. Crucially, it proceeds
with an understanding that many paths of technological development aspire for non-contested, business as usual, results, but fail
to realize this ambition as a result of reaching levels of growth that
engender deeply contested and surprising outcomes, some of which
fall so short of motivating intentions as to be best characterized
as perverse. When this happens, technologies originally developed
under a business as usual paradigm tend to get recast in a systems
engineering approach that attempts to minimize broader adverse
impacts.

One example is chloro-fluorocarbons (CFCs). Originally discovered by Thomas Midgley (who was nominated for, but not

awarded, a Nobel prize in chemistry in recognition of the discovery), CFCs were hailed has safe (i.e., nonexplosive), energy efficient, and non-toxic alternatives to problematic refrigerants such as ammonia or propane. The widespread adoption of CFCs enabled development of inexpensive refrigeration and air-conditioning technologies, with concomitant benefits in food preservation and the rapid growth of urban centers such as Atlanta and Phoenix in the American South. However discovery of CFCs in the atmosphere by James Lovelock (citation) soon led F. Sherwood Rowland and Mario J. Molina (1974) to hypothesize that unchecked use of CFCs would eventually lead to catalytic destruction of the stratospheric ozone layer. The response of the techno-industrial complex at first was complete denial and attempts to discredit Rowland and Molina. Nonetheless, CFCs were soon afterwards nearly completely banned by the Montreal Protocol in 1987. Both Rowland and Molina (along with Paul Crutzen) were eventually awarded the Nobel Prize in Chemistry that eluded Midgely. Now, the substitutes for CFCs (hydrofluorocarbons, or HFCs) are themselves implicated as significant contributors to an even more complex problem, global warming (Seager & Theis 2003, 2004).

Our main argument, therefore, is that while technological approaches to sustainability can be understood as pluralistic and evolve from one paradigm to another, sustainability science represents three major departures from other approaches: 1) it is predicated upon an understanding of the ethical requirements of technology development as extending beyond merely research or professional ethics into the domain of "macro" ethics (Allenby 2006); 2) it proceeds with an awareness of the necessity of consciously cultivating the interactional expertise necessary to carry out integrative (compared with reductionist) scientific research; and 3) it deliberately migrates from systems optimization to systems resilience perspectives (e.g., Seager and Korhonen 2010). This chapter addresses primarily the second point related to "interactional expertise," while the first regarding ethics is addressed by other sources (e.g., Allenby 2006, Seager & Selinger 2009), and the third regarding resilience is addressed in a companion publication (Mu et al., 2010). Consequently, ethics and resilience are mentioned only briefly here.

1. *Ethics.* Because sustainability science problems are also wicked problems, formulation of sustainability problems requires value judgments regarding boundaries, goals, and definitions that must be informed by some *ethical pre-positioning.*

Therefore, sustainability scientists need to develop the requisite skills that enable them to identify and directly address ethical issues that arise in relation to new technologies in the context of wicked problems. At a minimum, they need to be able to recognize behavioral and cognitive patterns of ethical significance, formulate ethical problems and employ the deliberative and moral reasoning skills necessary to work towards practical resolutions. However, the ethical reasoning skills required of a sustainability scientist transcend the professional ethics (such as the responsible conduct or research) necessary to responsibly carry out business as usual or systems engineering approaches. In sustainability science, problems of technology are embedded in complex, self-organizing systems. Therefore, ethical issues may emerge at a scale much larger than that of the individual. Where moral dilemmas are not traceable to the actions of individuals, or even to the collective action of an organization, profession or industry, they may result from the complex and dynamic *interaction* among many organizations and individuals. Consequently, they cannot be resolved without deliberation and collective action. The norms of professional ethics, which operate at the level of the individual decision-maker, are therefore inadequate to understand or identify ethical issues germane to wicked problems generally, and sustainability in particular. The sustainability scientist must acquire *sustainability ethics awareness* and the deliberation skills necessary to work through ethical issues in concert with others.

2. *Interactional Expertise.* Sustainability Science is transdisciplinary. Its hypotheses are formed by integrating environmental, social, and economic considerations with knowledge of a core science or engineering discipline. Therefore, sustainability scientists need to develop interactional expertise in disciplines that study subjects directly related to the triple bottom line of sustainable development, such as economics, public policy, and thermodynamics. To acquire interactional expertise, students must obtain in-depth and linguistically communicable understanding of the concepts, conventions, cognitive styles, and tacit knowledge that allow these disciplines to function as coherent paradigms and synthesize their newly acquired interactional expertises into a coherent *sustainability outlook.* Without this synthesis, it will be difficult for students to reliably conduct original

scientific research in fields like nanotechonogy, climate science, and bioengineering that accord with the principles of sustainable development.

3. *Resilience.* Sustainability has always confronted issues of risk. However, business as usual and systems engineering approaches to risk differ from those in sustainability science. Risk analysis in these approaches begins with hazard identification and strategies for risk mitigation may include armoring, redundancy, and resistance. By contrast, sustainability science borrows strategies from ecology such as diversity, adaptation, evolution, and renewal. These strategies are encompassed in an emerging approach to risk called *resilience*, which describes the ability of a system to respond to stressors without losing basic functionality or structure (Holling 1996). The most fundamental difference between the resilience strategies associated with sustainability science and other approaches is the realization that not all hazards can be identified in advance. As a consequence, the likelihood of catastrophic failure may be exacerbated by ignorance of hidden risks. By contrast, the resilience approach attempts to build flexibility and adaptability into systems that are responsive to any stressor.

2. Teaching Sustainability Science: Introducing a Pedagogy of Interactional Expertise

Graduate programs that prioritize business as usual and systems engineering approaches over sustainability science have the luxury of using established approaches to curricular development and teaching. By contrast, universities committed to sustainability science must accept the challenge of developing novel pedagogical tools. The common attempt to foster integrative education by enrolling students in a series of sophisticated introductory modules that cover a range of topics is inadequate for promoting the needed transdisciplinary training. For example, engineering students seeking broader contextual training in economics might register for an introductory graduate class such as "Microeconomics for Non-economists," or something similar, which would be taught by an economics professor, using pedagogical strategies drawn from economics. Such an approach often proves unsatisfying for both student and instructor alike. Typically, it remains clear to the instructor that even experienced engineering students are unable to solve advanced problems in economics after only one (or even two) introductory courses, while students sometimes object to assignments they perceive as outside their core expertise that they intuitively understand will not result in a marketable depth of knowledge. Using the common metaphor of the tree of knowledge, with different disciplines branching out from a common trunk or limbs, this traditional model requires the engineering students to "climb" back to the introductory limbs of economics without ever advancing out to the leaves. By contrast, we believe that a more appropriate mechanism entails using a *pedagogy of interactional expertise.* Returning to the metaphor, this pedagogy would devise strategies for teaching from leaf to leaf, rather than returning back to the base of the limb. Such a case would abandon a strategy of treating advanced engineering student like novice economists, and would instead focus on teaching engineers to *interact* with economists as experts, *without concern for the students' ability to practice economics.*

Our pedagogical inspiration comes from studies of interactional expertise (henceforth, IE). Over the last decade humanities and social science scholars have tried to establish two basic truths about IE: 1) it is central to developing *every form* of scientific expertise; and 2) it is used at the beginning of *all* genuinely interdisciplinary collaborations (Collins 2010; Collins and Evans 2007; Collins 2004; Gorman 2002; Collins, Evans, Gorman 2007).[1] "Interactional" and "contributory" expertises are different, though related, types of expertise. contributory experts are the class of professionals designated by the typical use of the word "expert." They develop specialist knowledge and skill through formal education and, in many cases, hands-on, experiential training and function at a recognized high level of ability. By contrast, interactional experts are not primary practitioners. They learn about a field, including its collective tacit knowledge, primarily by talking with the people who have acquired contributory expertise. The

---

[1] Some of the discussion of interactional expertise and related concepts appeared earlier in an NSF white paper, "Clarifying the Developmental and Pedagogical Dimensions of Interactional Expertise as a Function of Social and Psychological Relations Between Tacit and Explicit Knowledge" written by David Stone, Evan Selinger, Chris Schunn, and Barbara Koslowski for an National Science Foundation workshop called, "Acquiring and Using Interactional Expertise: Psychological, Sociological, and Philosophical Perspectives."

immersion enables interactional experts obtain considerable discursive expertise in specialized domains, even though they lack the practical skills required to make the contributions that directly advance the relevant professions.[2] Through this discursive prowess, interactional experts demonstrate they can see the world from a specialist's perspective –i.e., proffer authoritative technical judgments, make insider's jokes, and raise devil's advocate questions that revolve around ideas typically known only to specialists in a field.

Harry Collins, sociologist and godfather of the concept of IE, acknowledges that in practice, it can be difficult to prove one possess IE, particularly when confronted by skeptics who believe that only contributory experts can speak authoritatively about a field (Collins 2004b, 104-105). We recognize that given the constraints of graduate education, it is unrealistic to expect a course to transform sustainability science students from a range of disciplines into full-blown interactional experts, and that assessing interactional learning outcomes may be a challenge. A more realistic goal, albeit one appropriate for present purposes, is to create a new pedagogy infused with insights related to IE that could enable such students to become sufficiently "fluent" in expert discourse so as to be able to effectively interact outside their primary discipline. In such a context, IE would function as a regulative ideal.

Unfortunately, no one has created a successful pedagogy of

---

[2] Given the focus of this paper, it is only possible to present a brief summary of interactional expertise that is unable to convey the nuance found in scholarly literature and emerging conversations. For example, it is often pointed out that contributory experts typically possess interactional expertise. Otherwise they would not be able to make technical judgments in their fields that display knowledge of the underlying paradigm; nor, in the case of many sciences, would they be able to communicate with experts working within their broader specialties. Furthermore, in recent list-serv discussion the term "special interactional expert" has been used to emphasize the fact that contributory experts also develop interactional expertise. They do so in the sense that many disciplines are so specialized that in-between experimentalists and theorists exist a wide range of people whom we would think of as contributory experts. These contributory experts, in fact, have very little direct contact with either the practical matters involved in the experimentation or the complex mathematics involved in the theorizing, and so, in fact, derive most of their ongoing expertise from dialogue and conversation among their peers. Special interactional experts, then, designates the category of interactional experts who are completely "non-practice-based."

IE. Since IE involves mastering the *use* of language within a given domain, it uses tacit knowledge that cannot be fully explicated in terms of operational rules or through the application of formal knowledge. Because of its tacit dimension, IE is a skill that differs in kind from formal knowledge, and as such, it *cannot be acquired merely by reading texts, participating in lecture courses, or engaging with computer programs that provide portals into micro-worlds.* Indeed, ongoing feedback from contributory experts appears to be a prerequisite for obtaining IE. Furthermore, as Collins, who obtained IE in gravitational wave physics attests, acquiring IE is time-consuming. This means that attempting to obtain it could derail a promising career in a primary discipline. Therefore, the type of educational experience that is needed must make the defining features of IE accessible and inexpensive.

While questions remain about how best to impart the needed tacit knowledge, it is clear that the focus needs to be on accelerating "linguistic socialization," the process that Collins associates with acquiring IE. It thus may be sensible to adopt pedagogical strategies from foreign language instruction. These typically include techniques for obtaining: efficient memorization of words and phrases; grasp of semantics in written and oral forms of communication, including colloquialisms; knowledge of rules and cultural customs for structuring different types of conversational exchange; and, appreciation of relevant cultural and historical considerations.

To test this strategy, one of us (Seager) has recently experimented with this idea while teaching a graduate thermodynamics class. The class contained students from engineering, physics, mathematics, international studies, and climate science. Each student was aware of the fact that the class would be organized around the pedagogical techniques of foreign language instruction, and that the goal was to acquire interactional, rather than contributory expertise. As such, students were assigned to:

- Create a thermodynamic "alphabet" that described the mathematical variables typical in thermodynamic problems (such as V for volume, S for entropy, T for temperature and so on), their meanings and physical units.

- Build a lexicon of terms common in thermodynamics and a description of their multiple meanings. (For example, a popular mechanical engineering textbook defined temperature only as "the degree of hotness." The class concluded that a rigorous understanding of temperature was not necessarily required for solving problems of mechanical engineering, which was the focus of the text. Nevertheless, four other distinct definitions of temperature were culled from other sources).

- Complete an original research paper investigating cultural aspects of thermodynamics – for example, a biography of a prominent scientist in thermodynamics such as J. Willard Gibbs (the first student ever awarded the degree of Ph.D. in Engineering, who subsequently could not find a faculty appointment and never left New Haven, CT), James Joule (who "rigged" experiments to help popularize his formulation of the conservation law), or Ludwig Boltzmann (who was demonized by colleagues opposed to his statistical interpretation of entropy, suffered from depression, and eventually committed suicide even after his theories were widely accepted by others).

- Complete an immersion experience involving multiple problem sets in chemical engineering and mechanical engineering thermodynamics, culminating in a final project and exam.

Students expressed a wide range of reactions to the course. On the on hand, those familiar with the pedagogical approach (e.g., international studies) were comfortable and found much of the course accessible. Those from physical sciences (e.g., physics) were frustrated by the comparative lack of attention paid to complex problem sets, and the necessity of understanding cultural or societal context. (E.g., "I have trouble just looking at Gibbs the person, instead of focusing my paper on his work.") While preliminary, this experience may reinforce the idea that a course designed specifically to impart interactional expertise could be successful with students in disciplines outside the main subject area (in this case, thermodynamics), but be insufficiently attractive to students *within* the main subject area who are not previously attuned or self-selected. Nevertheless, where the primary goal is to speed the acquisition of interactional expertise for the purpose of working across disciplinary boundaries in sustainability science, it may prove advantageous to have both types of students working together in such a class.

Conclusion

Rittell and Weber formulated their original characterization of wicked problems while offering a critique of scientific expertise, which they deemed "doomed to failure" in the complex area of social planning. Similar criticisms have been levied against industrial-age science in the domains of ecology and economics, which, together with society, constitute the larger wicked problem identified as *sustainability*. Despite rapid adoption of the rhetoric of sustainability in nearly all science and technology disciplines, there has been very little effort at universities and research institutions to adapt the scientific enterprise to meet the challenge of sustainability as a wicked problem. To clarify some of the fundamental ways that sustainability training can be enhanced, we identified three principal shortcomings within the dominant scientific outlooks towards sustainability, "business as usual" and "systems engineering", that limit the conceptual resources available for creating innovation within sustainability education:

- A myopic focus on professional ethics to the neglect of macroethical issues germane to the complex interaction of many groups with multiple or competing views, and a lack of sensitivity to the ethical pre-positioning that comes from being embedded in the norms that typify different approaches to scientific and engineering research;

- A paucity of pedagogical techniques for teaching and incentivizing acquisition of the interactional expertise necessary to work effectively across disciplinary boundaries; and,

- A preoccupation with efficiency as the goal of high technology that undermines system adaptability, diversity and flexibility (i.e., resilience) in response to stressors or surprises.

While nascent attempts to foster a better understanding of what sustainability science is and has the potential to be exist in the ethics and resilience literatures, scant attention has been paid to developing techniques for training research scientists to acquire the IE appropriate for wicked problems. We proposed that pedagogical strategies borrowed from the foreign languages may speed development of interactional experts that further the preparation of a sustainability science community, and possibly even other science communities centered around different wicked problems.

To add support for this hypothesis, we discussed preliminary attempts to adopt such a strategy in a graduate thermodynamics course. This pilot project demonstrated that practical learning techniques can be devised that are consistent with this pedagogy, although their effectiveness in imparting IE remains an open question.

Bibliography

Allenby B. 2006. "Macroethical systems & Sustainability Science." *Sustainability Science* 1:7- 13.

Borgman, A. 1984. *Technology and The Character of Contemporary Life: A Philosophical Inquiry.* University of Chicago Press: Chicago IL.

Bruntland GH, Khalid M. 1987. *Our Common Future.* Oxford University Press: Oxford UK.

Collins H. 2010. *Tacit and Explicit Knowledge.* University of Chicago Press: Chicago IL.

Collins H. 2004. "How do you know you've alternated?" *Social Studies of Science* 34 (1):103 -106.

Collins H, Evans R. 2007. *Rethinking Expertise.* University of Chicago Press: Chicago IL.

Collins H, Evans R, Gorman M. 2007. "Trading zones and interactional expertise." *Studies in the History and Philosophy of Science* 38 (4): 657-666.

Clark WC, Dickson NM. 2003. "Sustainability Science: The emerging research program." *Proceedings of the National Academy of Sciences* 100(14):8059–8061.

Clark WC. 2007. "Sustainability Science: A room of its own." *Proceedings of the National Academy of Sciences* 104:1737-1738.

Connelly S. 2007. "Mapping sustainable development as an essentially contested concept." *Local Environment* 12 (3):259-278.

Davision A. 2001. *Technology and the Contested Meanings of Sustainability.* SUNY Press: Albany, NY.

Gorman M. 2002. "Levels of expertise and trading zones: A framework for multidisciplinary collaboration." *Social Studies of Science* 32(5):933-939.

Healy ML, Dahlben LJ, Isaacs JA. 2008. "Environmental assessment of single-walled carbon nanotube processes." *Journal of Industrial Ecology* 12 (3):376-393.

Heidegger M. 1977. "The Question concerning technology." In Krell DF (Ed.), *Martin Heidegger: Basic Writings*, 283-317. HarperCollins Publishers: San Francisco, CA.

Holling CS. 1996. "Engineering resilience vs. ecological resilience." In Schulze PC (Ed.) *Engineering within Ecological Constraints*. National Academy Press: Washington DC.

Iijima S. 1991. "Helical microtubules of graphitic carbon." *Nature* 354:56.

Kates RW, Clark WC, Corell R, Hall JM, Jaeger CC, Lowe I, McCarthy JJ, Schellnhuber HJ, Bolin B, Dickson NM, Faucheux S, Gallopin GC, Grübler A, Huntley B, Jäger J, Jodha NS, Kasperson RE, Mabogunje A, Matson P, Mooney H, Moore III B, O'Riordan T, Svedin U. 2001. "Sustainability Science." *Science.* 292 (5517):641 – 642.

Keefe R, Griffin JP, Graham JD. 2008. "The benefits and costs of new fuels and engines for light-duty vehicles in the United States." *Risk Analysis* 28 (5):1141-1154.

Komiyama H, Takeuchi K. 2006. "Sustainability Science: building a new discipline." *Sustainability Science* 1:1–6.

Michelcic JR, Crittenden JC, Small MJ, Shonnard DR, Hokanson DR, Zhang Q, Chen H, Sorby SA, James VU, Sutherland JW, Schnoor JL. 2003. "Sustainability Science and engineering: the emergence of a new metadiscipline." *Environmental Science and Technology* 37:5314-5324.

Mu D, Seager TP, Rao PSC, Park J, Zhao F,. 2010. "A resilience perspective on biofuels production." *Integrated Environmental Assessment Management.* Under review.

Norton BG. 2005. *Sustainability: A philosophy of adaptive ecosystem management.* University of Chicago Press: Chicago IL.

Oberdörster G, Oberdörster E, Oberdörster J. 2005. "Nanotoxicology: an emerging discipline evolving from studies of ultrafine particles." *Environmental Health Perspectives* 113(7): 823–839.

Oberdörster G, Stone V, Donaldson K. 2007. "Toxicology of nanoparticles: A historical perspective." *Nanotoxicology* 1 (1):2-25.

Rapport DJ. 2007. "Sustainability Science: an ecohealth perspective." *Sustainability Science* 2:77–84.

Rittell HWJ, Webber MM. 1973. "Dilemmas in a general theory of planning." *Policy Sciences* 4:155-169.

Rowland FS, Molina MJ. 1975. "Chlorofluoromethanes in the environment." *Reviews of Geophysics* 13 (1):1-35.

Seager TP, Selinger E. 2009. "Experiential teaching strategies for developing ethical reasoning skills relevant to sustainability." *Proceedings of the 2009 IEEE International Symposium on Sustainable Systems and Technology, Phoenix AZ, 18-22 May 2009.* Available online at http://ieeexplore.ieee.org/xpl/freeabs_all.jsp?tp=& arnumber= 5156721&isnumber= 5156678.

Seager TP, Theis TL. 2004. "A taxonomy of metrics for testing the industrial ecology hypotheses and application to design of freezer insulation." *Journal of Cleaner Production* 12:865–87.

Seager TP, Theis TL. 2003. "A thermodynamic basis for evaluating environmental policy trade offs." *Clean Technologies and Environmental Policy.* 4:217–226.

Simon JL. 1996. *The Ultimate Resource 2.* Princeton University Press: Princeton NJ.

Part III
Ethics & Politics of Expertise

◆

5

Feyerabend's Democratic Critique of Expertise

Introduction

Paul Feyerabend is perhaps best known for applying his descriptive philosophy of science to prescriptive questions about the proper function of science in the public sphere. Feyerabend's main reason for publishing *Against Method* is, he writes, "humanitarian, not intellectual" (2001, 3). His explicit goal is to "support people," not to "advance knowledge" (ibid.). Larry Laudan argues that Feyerabend's project is flawed precisely because it prioritizes a democratic political project over sound epistemological analysis. He notes that others with similar agendas have illicitly capitalized on Feyerabend's claim to have discredited the authority of scientific knowledge:

> Feminists, religious anthropologists (including "creation scientists"), counterculturalists, neoconservatives, and a host of other curious fellow-travelers have claimed to find crucial grist for their mills in, for instance, the avowed incommensurability and underdetermination of scientific theories. The displacement of the idea that facts and evidence matter by the idea that everything boils down to subjective interpretation and perspectives is–second only to American political campaigns–the most prominent manifestation of anti-intellectualism in our time. (Laudan 1990, x)

Feyerabend's politics are evident in his views on expertise. Feyerabend argues that experts ought to be regarded as public servants, and he insists that the exalted authority of the "expert" is incompatible with any legitimate democracy. Feyerabend resolutely declares that modern scientific experts have become *"ideologues"*:the more time and energy they devote to advancing a position, the more difficult it becomes for them to be open-minded to points of view that call their core beliefs into question.

Although Feyerabend legitimizes his position by appealing to speculative philosophy, his speculation supports only a modified version of his thesis and relies crucially on an exaggerated view of the epistemic capacities of laypeople. Feyerabend contends that laypeople ought to be the overseers of expert activities, not only to preserve democracy, but to enable experts to recognize how limited their point of view is. Conversely, he claims that if experts did not distort the value of their achievements, then laypeople would realize that they have more epistemological talent than experts give them credit for.

The problem is that Feyerabend *never explains what genuine lay talent is*. Instead, he uses poorly analyzed anecdotal evidence to produce a false account of who nonexperts are and what epistemic skills they possess. Because most of Feyerabend's critics focus on how he reduces experts to straw men, they overlook the fact that he romanticizes laypeople and their capacities. Laypeople are a disparate lot, with a variety of background skills, all of whom Feyerabend lumps together in a rhetorically effective but actually unjustified manner. This allows Feyerabend to imply that despite their apparent differences, all laypeople share a common aptitude for accurately criticizing expert advice. But the people whom Feyerabend designates as laypeople do not generally possess the skills, aptitudes, and attitudes that he attributes to them.

1. Scientific Experts as Ideologues

In its common usage, the word *ideology* designates either, first, a body of dogmas that reflect the particular perspective of an individual, group, class, or culture; or, second, a hidden agenda that is deliberately concealed in order for a person or group to maintain power over a deceived populace (or both of these meanings). Feyerabend means by "ideology" a manner of thinking that is characteristic of a group of people. He therefore uses *ideology* in the neutral first sense of the term, not in the politically devious second sense. If Feyerabend did not write in a polemical style, he could have made his point about ideological constraints on accurate perceptions of the world by using a more neutral phrase, such as "point of view" or "background theory."

"Ideologies" function according to what Feyerabend calls the "principle of tenacity," meaning that they have psychological origins and staying power. One does not become an ideologue by deliberately believing that certain arbitrarily selected texts, techniques, or teachers are sacred. Ideologies stem from being initiated into an inflexible view of the truth and from being trained in inflexible methods, beliefs, and practices. Feyerabend (1998, 62-65) argues that modern secular education confers psychological iner- tia on such dogmas by initiating students into a scientific prac- tice that tells them *how* to effectively achieve research objectives without providing the reasons *why* these goals are best achieved by proceeding in the orthodox manner. He sees this tendency as particularly detrimental in the natural sciences, where students are instructed in the technical dimensions of a field but are only minimally exposed to the historical arguments against the theor- ies that make the contemporary conventional wisdom seem true or useful (Feyerabend 2001, 11). Such education gives experts in the various fields the common characteristic of arrogance. They are overconfident about how to conduct research, and how to set the boundaries for generating accurate conclusions; as a result, they are prone to unreflectively dismissing alternative research methods and conclusions.

To qualify this claim, Feyerabend (1999, 112) notes that sci- entists have not always been dogmatic: "I think very highly of science, but I think very little of experts, although experts form about 95 percent or more of science today." He is neither critical of the institution of science in all of its actual or possible histor- ical manifestations, nor is he critical of all scientists. Feyerabend (1978, 284) is worried solely about the expertise accorded contem- porary "scientific" "experts":

> The way in which social problems, the problems of en- ergy distribution, ecology, education, care for the old and so on are "solved" in our societies can roughly be described in the following way. A problem arises. Nothing is done about it. People get concerned. Politi- cians broadcast their concern. Experts are called in. They develop theories and plans based on them. Power- groups with experts of their own effect various modi- fications until a watered down version is accepted and realized. The role of experts in this process has gradu- ally increased.

By focusing on how experts have gradually increased their power, Feyerabend distances himself from a variety of social critics. He rebukes Ibsen, Kropotkin, Levi-Strauss, and Marx for failing to distinguish between the type of science practiced during the seven- teenth and eighteenth centuries and the type practiced today. Ac- cording to Feyerabend (1978, 55-56), seventeenth-and eighteenth-

century science was an "instrument of liberation and enlighten-
ment" because "it restricted the influence of other ideologies and
gave the individual freedom for thought" (ibid., 75). By contrast,
he insists that in its nineteenth-and twentieth-century transformation, science "deteriorated" into a "stupid religion" (ibid., 55-56).

Science deteriorates, according to Feyerabend, because scientific
experts are no longer willing to put their theories to the test by
subjecting them to counter-hypotheses espoused by alternative
theoretical traditions. In order to defend this position, Feyerabend
appropriates John Stuart Mill's speculative account of how or-
thodoxy co-opts revolutionary views. Following Mill, Feyerabend
(1978, 30) argues that when a "new view is first proposed it faces
a hostile audience and excellent reasons are needed to gain for it
an even moderately fair hearing." No matter how good the reasons are that support the new view, they are often disregarded
because they conflict with widely accepted, traditional doctrines.
However, people in later generations are able to overcome the objections that a new view initially faces. Seen through younger,
fresher eyes, a view that was once rejected as unpersuasive becomes reassessed and studies are commissioned to evaluate its
merits. In some instances, such studies may raise difficulties for
the new view, but they can also lead to surprising successes for it.

As more and more people endorse the new view, consensus develops about its significance and it becomes formalized as a theory. As a theory, the new view becomes acceptable as a topic
for scientific discussion at conferences. As its acceptability grows,
diehards of the status quo feel an obligation to take the new theory
seriously and even make critical contributions to it. The theory
eventually becomes so widely accepted that its basic message is
transmitted beyond the esoteric confines of scientific meetings.
Ultimately, the theory enters the public domain through popular-
izations, such as introductory texts. "Unfortunately," Feyerabend
(1978, 30) writes, "this increase in importance is not accompanied
by a better understanding; the very opposite is the case." For

> problematic aspects which were originally introduced
> with the help of carefully constructed arguments now
> become basic principles; doubtful points turn into slo-
> gans; debates with opponents become standardized
> and quite unrealistic, for the opponents, having to ex-
> press themselves in terms which presuppose what they
> contest, seem to raise quibbles, or to misuse words. Al-
> ternatives are still employed but they no longer con-

tain realistic counter proposals; they only serve as a background for the splendor of the new theory.

2. The Losing Battle against Dogmatization

Support for Feyerabend's position can be found in the testimonial account of Allan Snyder. Snyder is the Director of the Centre for the Mind at the University of Sydney. He defines expertise as a "paradox." The paradox is that even though experts can display mastery in a field, they typically are less skilled than nonexperts at being able to locate detrimental prejudices that guide how their research is conducted. In his talk "Shedding Light on Creativity," which was delivered to the College of Physicians at Australian National University, he says:

> It is a curious fact that we are all blinded by our expertise. Those who have absolute mastery of a field are the very ones who find it hardest to question the foundations of the discipline....We are all experts at interpreting our visual world. The imposition of meaning on to the visual percept is something learned from birth. We are absolute masters at seeing what is needed of this world. But this very expertise exposes us to blatant prejudice in the form of illusions and deceptions that are so well known to psychologists. My particular research in this area concerns drawing. Don't you find it curious that none of us can draw natural scenes, unless of course we have been taught how to do so? Yet, certain brain deficient people can! And, these very people find it difficult to make sense of our visual world. They're not experts. And, how are we normal people taught to draw? We block our expertise by subterfuge. We take away the meaning of the object. For example, we look at something upside down or look at only a little piece of it at a time. In this way all of us can draw naturally. *So here's a paradox. Expertise seems to be associated with unavoidable blindness, but blindness can be overcome when there is no expertise.* (Snyder 1996, 710, emph. original.)

Snyder does not intend for this example to be understood as a condemnation of expertise. He is merely trying to illustrate that expertise is *simultaneously an enabling and disabling phenomenon.*

Expertise is enabling because it allows people to make decisions "rapidly" and even "automatically" (Snyder 1996,710). But it is disabling in that, as an expert becomes habituated in a fixed mode of thinking, she finds it increasingly difficult to re-examine or question the foundations of her beliefs. Further evidence for the position that ideology effects experts comes from David Sackett, a clinical-trials physician who started the Department of Clinical Epidemiology and Biostatistics at McMaster University and the Center for Evidence-Based Medicine at Oxford. Sackett defines experts as people who commit "two sins." The first sin of experts is that they grant the opinions voiced by other experts "far greater persuasive power than they deserve on scientific grounds alone" due to deference, fear, and respect (Sackett 2000, 1283). The second sin relates to the first:

> The second sin of expertness is committed on grant applications and manuscripts that challenge the current expert consensus. Reviewers face the unavoidable temptation to accept or reject new evidence and ideas, not on the basis of their scientific merit, but on the extent to which they agree or disagree with the public positions taken by experts on these matters. (Ibid.)

Sackett was originally an expert on the subject of patients' compliance with therapeutic regimens, but when he realized the two sins he was committing, he "wrote a paper calling for the compulsory retirement of experts and never again lectured, wrote, or refereed anything to do with compliance" (Sackett 2000, 1283). Sackett not only believes that his expert status was unduly influencing the way research in therapeutic regimens was conducted, but also that his influence could not be adequately monitored or corrected. And he interprets the excessive influence he possessed as an expert not as an isolated problem but as a general one. He claims that by virtue of their authority, experts thwart the acceptance of new ideas and slow the progress towards truth.

After following his own advice and retiring from work on compliance, Sackett switched fields, applied his methodological skills to new research, and eventually became an expert in evidence-based medicine. Fearing that his renewed expert status was again negatively influencing how research was being conducted, because his conclusions were given too much credence and his opinions too much weight-a technical term was invented that bears his name, "Sackettization"-he decided to retire once more and "never again lecture, write, or referee anything to do with evidence based clinical practice" (ibid.). Today, Sackett focuses his energies on thinking, teaching, and writing about issues related to a third field of interest, randomized trials.

Although the evidence they present is only testimonial, both Sackett and Snyder give us some basis for accepting at least a modified version of Feyerabend's position: if their accounts are representative, it seems that the longer an expert devotes to advancing a given perspective, the more difficult it becomes for her to be open-minded to viewpoints that call this perspective into question. Even good-faith experts find it difficult to be adequately self-reflective. They believe that they are aware of their limitations and that they are willing to admit when they cannot solve a problem or answer a question. But in reality, according to Feyerabend, while modern scientific experts can admit that people from other disciplines can better solve problems external to their own field, they find it difficult to concede that they could be incorrectly pursuing a problem (or pursuing a non-problem) that happens to be central to their field.

3. The People vs. Their Experts

Feyerabend goes astray when he tries to make a grander claim than the modified skepticism about expertise just outlined. But he justifies his excessively bold claim by turning to an examination not of experts but of their opposites: "laymen." Feyerabend (1978, 96-97) claims that one way to break the hold of expert ideology is to empower nonexperts through institutional provisions that allow for laypeople to judge expert viewpoints and research agendas:

> Duly elected committees of laymen must examine whether the theory of evolution is really as well established as biologists want us to believe, whether being established in their sense settles the matter, and whether it should replace other views in school. They must examine the safety of all nuclear reactors in each individual case and must be given access to *all* the relevant information. They must examine whether scientific medicine deserves the position of theoretical authority, access to funds, privileges and mutilation it enjoys today. ...The committees must also examine whether people's minds are properly judged by psychological tests, what is to be said about prison reforms and so forth.

Like many of Feyerabend's positions, his view that laypeople should have absolute regulatory control over expert activities is overstated. By presenting us with a lay/expert dichotomy, Feyerabend leads the reader to believe that only two choices exist: either that experts should be *absolutely* free to attend to their research without constraints, or that their judgments should *always* be trumped if laypeople disagree with it.

But why should expert judgments *ever* be overridden by lay points of view? Feyerabend presents two answers to this question. First, if experts paid more attention to how nonexperts learn, then they might realize that their own authority is overly esteemed. Feyerabend tries to prove this point by discussing the acquisition of medical skills. Second, if experts became sensitive to lay criticism, then they would realize how easy it is for experts to overlook basic gaps in their arguments. Feyerabend tries to prove this point by discussing the behavior of expert witnesses under legal cross-examination.

Feyerabend (1987, 307) tries to establish the salutary effect of lay oversight in reducing experts' arrogance by arguing that one important reason that nonexperts deem experts authoritative is that experts exaggerate how difficult it is to cultivate their talent. But modern science, Feyerabend argues,

> is not at all as difficult and as perfect as scientific propaganda wants us to believe. A subject such as medicine, or physics, or biology appears difficult only because it is taught badly, because the standard instructions are full of redundant material, and because they start too late in life. During the war, when the American Army needed physicians within a very short time, it was suddenly possible to reduce medical instruction to half a year (the corresponding instruction manuals have disappeared long ago, however). Science may be simplified during the war. In peacetime the prestige of science demands greater complication.

Feyerabend suggests three things in this passage. First, if laypeople were not misled by distorted accounts of how to acquire expert skills, they might realize that they could become experts in technical fields like medicine faster than is generally believed. Second, certain circumstances are more conducive than others to allowing experts to exaggerate how difficult it is to acquire their expertise. Third, experts will exaggerate their expertise if doing so increases their prestige. (Here, Feyerabend seems to risk veering into the second, more conspiratorial view of ideology.)

These implications seem to follow only given the following implicit argument. Although four or more years of training in medical school are typically required for students to become physicians, medical instruction was given to army-trained physicians in half a year during wartime. From the contrast between these two lengths of time, Feyerabend concludes (1) that standard medical training is inefficient, and (2) that doctors would be less esteemed were medical training more efficient. What is more, because they believe (2), experts conspire to keep (1) from becoming common knowledge.

Feyerabend thus makes two contentious assumptions: (A) that both medical school and army-trained physicians have the same set of skills, such that graduates of both types of training program are equipped to handle the same spectrum of medical procedures; and (B) that accelerated army training and conventional medical-school training produce physicians with comparable short-and long-term error rates.

Since Feyerabend claims that the instruction manual for army training has disappeared, it is unclear how accurate (A) and (B) are. Regarding (A), it seems likely that the types of skill obtained in the army training program were selective; presumably they mainly prepared recruits to handle the aspects of medicine relevant to wartime, such as treating burns. By contrast, physicians trained in medical school probably have a greater general understanding of medicine and can respond to more medical problems than could physicians trained in the army's accelerated program. Regarding (B), Feyerabend never establishes whether the techniques used by the army-trained physicians were as safe for patients in the long term as the techniques that would be used by physicians trained in medical school.

Furthermore, Feyerabend overlooks the possible connection between affect and the acquisition of expertise. According to Hubert Dreyfus (1986), if a student does not feel emotionally connected to the successes and failures that result from acting according to personally chosen strategies, she will be less likely to acquire expert-level skills. This point is relevant to Feyerabend's example because during wartime, medical trainees might well feel more passionately about their education than medical students do in non-wartime conditions; they are liable to be confronted with life-and-death situations involving their comrades in arms, often their friends. But if cadets do indeed feel more committed than their medical-school counterparts, and if passionate commitment can accelerate the learning process, then even if (A) and (B) are correct, Feyerabend's conclusion is still unsupported.

The bulk of Feyerabend's case against experts consists of similarly unfounded arguments for the potential expertise of laypeople. Consider these three passages:

> Conceited and intimidating scholars, covered with honorary degrees, university chairs, presidents of scientific societies are tripped up by a lawyer who has the talent to look through the most impressive piece of jargon and to expose the uncertainty, indefiniteness, the monumental ignorance behind the most dazzling display of omniscience: science is not beyond the reach of the natural shrewdness of the human race. (Feyerabend 1978,98, emph. original)

> That the errors of specialists can be discovered by ordinary people provided they are prepared to "do some hard work" is the basic assumption of trial by jury. The law demands that experts be cross examined and that their testimony be subjected to the judgment of a jury. (Ibid.,97)

> One of the most exhilarating experiences is to see how a lawyer, who is a layman, can find holes in the testimony, the technical testimony, of the most advanced expert and thus prepare the jury for its verdict. (Feyerabend 1998,61)

In short, laypeople have a general capacity for detecting and unmasking logical fallacies in experts' arguments.The problem with the example of the lawyer developed in the first and third passages, however, is that it neglects the differences between laypeople and lawyers. A lawyer traditionally becomes a skilled crossexaminer not simply because she is willing to "do some hard work," but because she possesses a formally taught ability (obtained through training in law school, mock-trial practice, and courtroom experience) to dissect and refute witnesses' testimony. In treating a lawyer's cross-examination of an expert witness during a trial as a proxy for the abilities of the average layperson to criticize a scientific expert, Feyerabend also overlooks the fact that the aim of a lawyer is not the establishment of scientific truth, but victory in an adversarial legal setting. The same lawyer who may be able to expose fallacies proffered during expert legal testimony, may, when engaging with experts outside of the courtroom (e.g., *qua* patient with a physician), be less talented in exposing fallacious arguments. A lawyer who attempted to cross-examine his doctor (if the doctor allowed it) might succeed only in embarrassing the "witness"—not necessarily in achieving a more accurate diagnosis.

The example of the juror, taken together with the example of the lawyer, is supposed to justify the sweeping claim in Feyerabend's first passage about the accessibility of scientific truth to laypeople. But in reality, jurors do not directly engage with expert witnesses; they simply listen as some experts (scientists) are cross-examined by others (experts in the art of discrediting scientists rhetorically). And when jurors do take on an active role, they prove to be quite incapable of understanding the implications of scientific testimony, or even of performing simple tasks of memory and logic. In punitive-damages cases, for example, a recent study reporting on elaborate experimental data on jury deliberations found that

> only 5 percent of the jurors ... remembered and understood the judge's instructions setting out the legal standard for their task, and this ignorance held even when the jurors were provided with memory aids, including various forms of written reminders ....

> Far from making punitive damages awards more rational, [jury] deliberation increased their severity and unpredictability. Moreover, despite painstaking instructions and warnings, civil jurors consistently fell prey to "hindsight bias," by which they saw any accident that did happen as an accident that was waiting to happen. So, for example, when participants in these studies were presented with a fact scenario concerning the operation of a railroad and were asked if they would approve of its continuing operation without further safety precautions, two-thirds of them said "yes." But when they were presented with precisely the same facts, and were then told that an accident had occurred, the same proportion had little trouble concluding that the railroad's decision to operate without taking further precautions had been reckless and therefore warranted the award of punitive damages. Jurors have an ironclad aversion to the use of a formal calculus to make decisions about safety procedures. Any

time evidence came to light that a company had performed a cost-benefit analysis in deciding what safety precautions to take in a given circumstance, the level of punitive damages awarded spiked upwards .... Cost-benefit analyses are said to demonstrate how depraved corporations "put a price tag on human life."... However, neither businesses nor individuals can function without "putting a price tag on human life." Each time we get into a car, step onto an airplane, swim in the ocean, or head to the city, we put a price on human life; we take a risk after either calculating or assuming that it's worth it. To operate in the real world, businesses are forced to make similar calculations concerning the lives of others. If something happens to go wrong, however, the willingness of companies to calculate these risks explicitly is taken by juries as a sign of the moral degeneracy of corporate America.

To add insult to injury in these matters ...the higher the value that a company places on a human life in its cost-benefit calculations, the higher the damages [assessed against the company]. The culprit here is the "anchoring effect" by which the jurors, at sea in their task of attaching a dollar amount to the alleged transgression, grabbed pretty much any number at hand as the base point for their calculations. These numbers include the (legally irrelevant) amount asked for by the plaintiff's attorney (the more he asked for, the more he got), and the (legally irrelevant) level of compensatory damages, as well as the valuation of human life drawn from the company's internal cost-benefit analysis. (Kersch 2003, 124-2 5.)

None of this experimentally derived information, of course, was available to Feyerabend, and conceivably jurors in real-world trials, for Dreyfusian reasons, may suddenly acquire better memories and reasoning faculties than jurors in laboratory experiments. But the experimental evidence is *prima facie* plausible, and Feyerabend provides no evidence at all to suggest that, on the contrary, jurors tend to be logical, let alone that they tend to be as knowledgeable as experts. What he does instead is treat a democratic political ideal-that of the competent juror-as if it were an empirical reality: "That the errors of specialists can be discovered by ordinary people provided they are prepared to 'do some hard work' is the basic *assumption* of trial by jury" (Feyeraband 1978, 97, emph. added). But we are given no reason to accept this assumption as having any basis in reality.

His politics may also drive Feyerabend's dichotomization of the world into experts (elites) and laypeople (the people). In any event, it is clear that this dichotomy is what enables him to lump jurors and lawyers together as laypeople and to credit them with the attributes of experts; and it is upon this move that his case against expertise largely rests.

4. If These Aren't Experts, Who Is?

Feyerabend fails to recognize that there are many different ways one can be outside of scientific practice. For example, someone who is *uninformed* about physics might be completely ignorant of even its basic principles. A *"well informed amateur"* (Ihde 1998, 134-35) might be ignorant of many aspects of physics, but familiar with a particular problem area. A *frontier researcher* might know quite a bit about physics, yet have views on some aspect of it that differ greatly from those espoused by the dominant community of physicists. And an *absolute outsider* might oppose any claims that are justified by physics, on the basis of an alternative worldview.In the following illustrative passage, Feyerabend (1978, 88-89) blends these four different senses of being a scientific outsider together. Scientific mistakes that have been hidden by the experts' consensus view, he writes,

> *can be* discovered by laymen and dilettantes, and often *have been* discovered by them. Inventors built "impossible" machines and made "impossible" discoveries. Science was advanced by outsiders, or by scientists with an unusual background. Einstein, Bohr, Born were dilettantes and said so on a number of occasions. Schliemann who refuted the idea that myth and legend have no factual content started as a successful businessman, Alexander Marshack who refuted the idea that Stone Age Man was incapable of thought was a journalist....Columbus had no university training and learned Latin late in his life .... The Chinese communists of the Fifties who forced traditional medicine back into the universities and thereby started most interesting lines of research the world over had only little knowledge of the intricacies of scientific medicine. How is this possible? How is it possible that the ignorant

or ill-informed can occasionally do better than those who knew their subject inside out?

Feyerabend tries to convince the reader that in some sense, Einstein, Bohr, Schliemann, Marshack, Columbus, and practitioners of traditional Chinese medicine are all "laymen," equally classifiable as "ignorant" and "ill informed," equally outsiders to expert culture. Yet all that he has really shown is that sometimes people who begin their lives engaged in one career are eventually able to make contributions to a field that they were not initially expert in. This passage (and others like it that appear throughout Feyerabend's work), fails to establish that all of these initial outsiders remained outsiders the fields they eventually made contributions in. Classifying Einstein and Bohr as dilettantes, at least in part on the basis of them approving of being classified in this manner, is naïve; it is comparable to believing that James Watson (1969) was really as awkward as he depicts himself being during the relevant period leading up to the discovery of DNA. This depiction is not historically accurate. It is a personae that Watson dons in order to make readers even more impressed with his accomplishments than they would be on scientific grounds alone.

What is extremely valuable in Feyerabend is the insight that expertise can simultaneously be an enabling and disabling phenomenon. But while criticism may help an expert become less attached to an ideological perspective, there is nothing about *lay* criticism that especially qualifies for this role. In his attempt to demythologize all modern scientific experts, Feyerabend idealizes the lay perspective as a general cognitive framework and thereby commits the "superman fallacy." Whereas the straw-man fallacy entails presenting a caricatured depiction of a rival's position, the "superman fallacy" entails idealizing the protagonist one wishes to defend—in Feyerabend's case, "the people"—with super-powers that are not actually present. Laudan thus seems justified in being suspicious about the relation between politics and epistemology in Feyerabend's work.

Bibliography

Dreyfus, Hubert and Stuart Dreyfus. 1986. *Mind Over Machine: The Power of Human Intuition and Expertise in the Era of the Computer*. New York: Free Press.

Laudan, Larry. 1990. *Science and Relativism*. Chicago: University of Chicago Press.

Feyerabend, Paul. 1978. *Science in a Free Society*. London: Verso.

Feyerabend, Paul. 1998. "How to Defend Society Against Science." In *Introductory Readings in the Philosophy of Science*, ed. E.D. Klemke, Robert Hollinger, and David Wyss Rudge. New York: Prometheus Books.

Feyerabend, Paul. 1999. "Experts in a Free Society." In *Knowledge, Science and Relativism, Philosophical Papers,* vol.3, ed. John Preston. Cambridge: Cambridge University Press.

Feyerabend, Paul. 1975. *Against Method*. London: New London Books.

Feyerabend, Paul. 2001. *Against Method*. 3rd Edition. New York: Verso.

Jasanoff, Shelia. 1995. *Science at the Bar: Law, Science, and Technology in America*. Cambridge: Harvard University Press.

Sackett, David L. 2000. "The Sins of Expertness and a Proposal for Redemption." *British Medical Journal* 320: 1283.

Selinger, Evan and Robert Crease. 2002. "Dreyfus on Expertise: The Limits of Phenomenological Analysis." *Continental Philosophy Review* 35: 245-279.

Snyder, Allan. 1996. "Shedding light on creativity." *The Australian and New Zealand Journal of Medicine* 26 (1996): 709-711.

Watson, James. 1969. *The Double Helix: A Personal Account of the Discovery of the Structure of DNA*. New York: First Mentor.

6

Catastrophe ethics and activist Speech: Reflections on Moral Norms, Advocacy, and Technical Judgment

Written with Paul Thompson and Harry Collins

Introduction

In what follows critical examinations is given to the issue of whether there are ethical dimensions to the way that expertise, knowledge claims, and expressions of skepticism intersect on technical matters that influence public policy, especially during times of catastrophe. Rather than offering a single judgment about ethics and expertise, we intend to enhance discussion of a situation that will be defined as "Catastrophe ethics" through comparison of two views: (1) the philosophical perspective on appropriate appeals to expertise in public sphere discourse that combines insights from discourse and virtue ethics, and (2) the so-called third wave sociological perspective on expertise that construes activists as having normative obligations different from those of political leaders and the press.[1]

---

[1] To represent contrasting views fairly, this essay has three authors: Harry Collins is a founding principal of the second wave of science and technology studies, and an exponent of the Cardiff third wave approach to expertise; Paul Thompson is a biotechnology ethicist; and Evan Selinger is a philosopher of technology and science who initiated the present discussion, framed its parameters and structure, and regularly played the role of interdisciplinary translator. To avoid potential confusion concerning authorial voice, it should be noted that Collins conveys the "sociological rejoinder" (Section 6) in his own terms, while Selinger has provided much of the reconstruction of the philosophical view, drawing from earlier works by Thompson and extended recent conversations. Disparities between how Thompson's views were presented earlier and how they are presented here can be attributed to this process. In an attempt to provide scholarly analysis, the essay uses third-person narration to convey points that can be directly attributed to the authors themselves.

The discussion proceeds as follows. First, we clarify the relationship between expertise and public testimony. Second, we introduce the concept of "Catastrophe ethics" and explain why some theorists believe that it designates a unique moral situation. Third, we present an overview of a controversy that currently is being framed as a Catastrophe ethics scenario—the matter of African nations rejecting genetically modified (GM) food. Fourth, we assess a contentious aspect of the GM debate. Specifically, we offer a deflationary counter to Robert Paarlberg's charge in *Starved for Science* (2008) that anti-GM activists are behaving immorally. We conclude by addressing the question of whether activists should ever be morally accountable for how they convey their views about technical matters, and we offer two different answers for consideration. Endorsing the combination of virtue and discourse ethics leads one to conclude that there is at least one instance where activists who dissent from the core group of scientific experts are morally obligated to frame their case to the public with a robust set of qualifiers that effectively undermines the persuasiveness of their case. By contrast, the sociological commitment to the primacy of intellectual freedom—associated with the social constructivism of the so-called second wave of science studies—leads to the opposite conclusion. It denies that any cases exist or could exist that justify holding activists who are not deliberately deceiving others morally accountable for how they present their positions on technical matters in the public sphere. Rather than arguing that one approach is better than the other, we conclude by explaining how others can appeal to the juxtaposition we presented between philosophy and sociology to advance the current discipline-based discussions of expertise and normativity.

1. Expertise and Public Testimony

The term "expertise" has been used in several overlapping senses. Literature on risk communication and structured decision-making sometimes classifies anyone who has information that is relevant to a decision as having expertise. Here, the extremely broad construal of expertise seems to have been developed in response to documented failures in risk management and environmental decision making where decision makers or analysts with scientific credentials failed to consider facts that might have been common knowledge to local practitioners. Brian Wynne's (2006) sociological work on Cumbrian sheep farmers is a paradigm case of such work. The motivation was to broaden the notion of expertise beyond that of those holding academic credentials. But a conceptualization that emphasizes decision relevance neglects considerations that are traditionally tied to the concept of expertise. Some individuals may have information that is relevant to a decision simply in virtue of having been placed in a given situation at a particular time. Eyewitness testimony exemplifies this kind of knowledge. In contrast, the concept of expertise typically implies proficiency in a domain of practice so that the testimony of an acknowledged expert will have priority over that of a lay witness in matters where proficiency in the practice might inform perception or judgment. Thus, Wynne's sheep farmers would certainly qualify as experts on sheep farming, and their testimony on how regulations might affect sheep farming should be given priority over testimony of an atmospheric scientist or radiation specialist. But the testimony of an eyewitness to an event might be regarded as that of a valid informant without also being regarded as that of an expert.

It is perhaps incumbent upon all persons offering testimony that would be relevant to public decision making of the sort implied by risk management and environmental policy cases to speak truthfully. Thus, an eyewitness can be said to be under a moral (and often legal) duty to speak truthfully when offering testimony, and presumably the same goes for experts offering testimony. The point in question in this article concerns whether the concept of expertise implies moral responsibilities that amplify those of any person called upon to give public testimony.

Many codes of professional conduct require practitioners to show respect for the parties they counsel. This sense of respect can be derived from a variety of principles and intuitions. It requires advocates to ensure that their behavior accords with the following two obligations: (1) do not distort facts to make a case seem more persuasive than it actually is, and (2) carefully consider the implications that are likely to follow if the position one advocates for is accepted by the parties one is trying to convince.[2]

Given the number of professions that share this view of expertise, the justification for respect appears to run deeper than the norms that define any particular field. For similar reasons,

---

[2] Of course, it would be a mistake to take an unduly simplistic approach to the relation between advocacy and distortion. For example, physicians may occasionally lie to patients to make them feel better, and medical professionals of varying kinds may even routinely lie to family members to promote the greater good. In a more fundamental sense, we can say that given routine use of placebos, lying is even a constitutive part of medical expertise.

an advocate's obligation to be respectful is not reducible to social etiquette. Violations of etiquette invite the rebuke that one should not have broken the conventional rules of proper behavior. However, because etiquette is a matter of "precepts" that different communities can define according to "disparate and apparently arbitrary codes" of practice, it is compatible with relativism (Scapp and Seitz 2006, 2).

Professionals who offer expert advice to clients are *morally obligated* to respect their audiences because the primary purposes of advocacy are: (1) to persuade others to act in accordance with one's advice on a given issue, and (2) to convince others to accept the consequences that follow logically and causally from such acceptance. Members of a profession thus aspire to be cognitive authorities, and they bear responsibility for attempting to influence people to conduct their affairs in the particular ways that they deem justified.

When expert advice is offered appropriately, it can enhance decision making, even when a specific relationship to a client has not been formally established. For example, responsible advocates of global justice offer the public in first-world countries a valuable service by bringing to their attention issues of far-away suffering and misery that the mainstream media ignores or downplays. These advocates do have an agenda—to get the public to direct resources to those who are suffering—but that agenda can be justified by virtue of how it accords with widely shared convictions about justice and beneficence. Crucially, the advocate typically only ask their audiences for modest monetary support, and do so through value-laden but noncoercive forms of communication, such as print advertising and mailing campaigns, that are easy to reject.

Although one may question whether advocates for global justice possess expertise in the relevant sense, there are many cases where only persons having some degree of expertise would be in a position to advocate in the sense alluded to above. Awareness of the carcinogenic properties of tobacco smoke, the toxic properties of agricultural chemicals, and the long-term impact of emitting greenhouse gases would not have occurred were it not for public testimony offered by persons having expertise in scientific disciplines that made them aware of these hazards long before any layperson. Thus one possible expansion of the expert's moral responsibility may lie in the area of a fiduciary responsibility that must be exercised in situations where one's expertise makes one particularly qualified to offer testimony on matters of public interest.

The uniqueness of this responsibility should not be overstated, however. Someone possessing knowledge relevant to a case, such as an eyewitness, would also be thought to have a responsibility to come forward and offer relevant testimony, subject to *ceteris paribus* criteria that need to be elaborated in the present context. Yet it still seems plausible to think that proficiency in a technical field may introduce fiduciary responsibilities that extend more broadly than those of a person whose relevant knowledge has nothing to do with expertise. There is little reason for others to place particular confidence in the layperson's ability to perceive and report observations reliably, for example. The point can be illustrated by considering the abuse of expertise.[3]

When advocacy for a given position or view is abused, it can engender adversity and violate trust. While different theories offer different reasons for construing the violation of trust as a moral harm (e.g., Immanuel Kant's deontological philosophy condemns the intent to make a lying promise on the grounds that such a goal threatens very the institution of promising), the virtue ethics perspective is especially strict. For the virtue ethicist, an advocate who commits duplicitous actions is best conceived of as a person who is likely to have been influenced by blameworthy character traits and dispositions. The justification for this attribution lies in the fact that virtue ethics—at least the Aristotelian variety—understands character traits to be virtues and vices that one develops through habituation and that are manifested in habitual action. What is more, because the development of character traits is seen to follow from the milieu in which one works and the people whose company one keeps, attributions of character in the Aristotelian tradition have a social dimension.

Those who violate trust thus are construed as people who have a penchant for dishonesty; they are likely to violate trust often. But more than this, to the extent that these character traits are seen as representing a group, it becomes reasonable to presume that the traits are typical of the entire group, in as much as virtues or vices are presumed to be reinforced and habituated through socialization. Thus abuse of a trusted position is doubly problem-

---

[3] For an authoritative historical treatment of who counts as a reliable scientific witness, see Shapin 1995.

atic when the trust accorded to a given individual is based upon that individual's ability to represent a community of practice. All practitioners in the community are, to some degree, tarnished.

The mundane case of used car sales exemplifies this point about virtue and trust (Thompson 1999). Selling used cars is widely thought of as a deceptive profession. Although an individual used car salesperson may be an exception to the rule and exhibit virtue, and although unscrupulous used car salespeople may have lots of automotive knowledge and skill, the social fact remains that the profession as a whole is tarnished. Any reasonable consumer looking to purchase a car will exhibit a high degree of skepticism toward all used car salespeople. Such skepticism effectively disqualifies any particular used car salesperson's claim to put expertise in the service of trustworthy advocacy.

Thus to the extent that testimony, advocacy, or advice is made based upon proficiencies that depend upon an individual's ability to contribute to a community of practice (that is to say, when testimony is based on expertise), it is at least plausible to assert that the person offering this testimony has responsibilities not only to a particular client but also to other members of the community of practice, as well as to the public at large. These can be called fiduciary responsibilities in that they involve (1) ways in which the body of knowledge or practice can contribute to public benefit and to the avoidance of harm, and (2) it becomes important to maintain confidence in this body of knowledge and, by extension, in those who represent themselves as being proficient in it. A general theory of experts' ethical responsibilities (if any) must wait. In what follows, we concentrate on scenarios in which the need for action on a timely basis is critical, and in which the potential consequences of action (or the lack thereof) are high.

2. Catastrophe ethics

Paul Krugman, the Nobel Prize–winning professor of economics and international affairs at Princeton University, gave paradigmatic expression to the catastrophe ethics position in his *New York Times* editorial "Can This Planet Be Saved?" (2008). In this brief piece motivated by the offshore drilling controversy, Krugman presents the policy choices that follow from the global climate change debate in terms of an updated version of Pascal's Wager:

> It's true that scientists don't know exactly how much world temperatures will rise if we persist with business as usual. But that uncertainty is actually what makes

action so urgent. While there's a chance that we'll act against global warming only to find that the danger was overstated, there's also a chance that we'll fail to act only to find that the results of inaction were catastrophic. Which risk would you rather run?

For Krugman, citizens and policy makers alike are faced with a clear binary. Either global warming is an enormous problem with high ecological, social, and political stakes or it does not present a high risk of bringing about catastrophe. If the former, then (depending on one's conception of intergenerational justice) the prudent, rational, and perhaps even moral course of action is to do everything possible to mitigate against deleterious effects. If the latter, then the resources put toward mitigating its presumed deleterious effects may appear to be poorly allocated, at least retrospectively.[4]

To settle the question of risk, Krugman appeals to an unnamed study of a "wide range of climate models" that he tells us was performed by Martin Weitzman, a Harvard economist. In light of the dismal conclusions that Weitzman is said to have reached, Krugman contends: "It's sheer irresponsibility not to do whatever we can to eliminate the threat." Such a by-any-means-necessary approach to environmentalism entails branding those who dissent from the Weitzman view of global climate change as irresponsible citizens: "The only way we're going to get action (...) is if those who stand in the way of action come to be perceived as not just wrong but immoral." While society can accept technical disagreement during routine, science-as-usual conditions, Krugman apparently believes that it should not tolerate dissent during times of catastrophe ethics. Had Krugman meant to articulate a less radical thesis—perhaps something like "During times when technical debate has profound public policy implications, severe legal penalties should be imposed on hired guns who willingly falsify data"—he certainly would have been capable of conveying it.

Whereas Krugman directs his moral outrage at everyone who disagrees with his accepted view of global climate change, others involved in catastrophe ethics discussions restrict their condem-

---

[4] This way of putting the global warming wager brackets the issue famously raised by Bjorn Lomborg over whether the proposed cure is, in fact, more catastrophic than the disease.

nation to activist organizations and government officials. For example, during a 2002 crisis, Zambia's President Levy Mwanawasa rejected offered donations of GM food, claiming that the aid was "poisonous." Since such a decision appeared to contradict the consensus judgment of a core group of scientists, Guy Scott—a former Zambian minister of agriculture—condemned the activist organizations that validated the president's rejection:

> What we will see now is how many people die as a result of the disruption of the relief programme—and how the various international NGOs that have spoken approvingly of the government's action will square the body count with their various consciences. (qtd. Carroll 2002)

To imply that Greenpeace and Friends of the Earth should have troubled consciences because their advice is partly responsible for unnecessary deaths is to say that technical judgment can be assessed in moral terms under the conditions of catastrophe ethics.

The U.K. newspaper *The Independent* (2008) rebuked government officials in Africa for their poor response to the GM controversy:

> If we fail to increase agricultural yields on our planet, people will starve. Faced with such a chilling scenario, it would be positively immoral for governments to reject GM out of hand.

Many have argued for this position, and as the following claim made by Martin Makinede—former dean of the Department of Animal Science at the University of Venda for Science and Technology in South Africa—illustrates, it is a view that links conceptions of scientific objectivity and technical risk assessment with a conception of human flourishing.

> The economic benefits of GM crops are certainly tangible, and Africa's poorest farmers should have access to this technology. If the developing world's legitimate demands regarding food production are not attended to, there will be dire consequences for both developing and developed worlds. . . . For the benefit of humankind, we must end the squabbling over biotechnology and allow objectivity to prevail. Scientists are developing strains of rice, cassava, and other staple foods that

are fortified with more nutrients. Crops are also being developed to generate their own protection against
pests and disease. Of course there may be potential
risks. But the benefits and the severity of the need
make it clear that sensible approach is to minimize
the risks, not to abandon biotechnology. (Makinede
2004, 125)

Views like the ones just listed are becoming commonplace responses to nonconsensus advocate judgments on the technical
dimensions of the global climate change and GM food debates.
Moreover, as the power of science and technology expands, we
can expect other issues to become subject to similar framing. At
least implicitly, partisans in the global climate change debate and
the GM debate justify applying the catastrophe ethics framework
to their arguments by appealing to the following utilitarian considerations and counterfactual narratives. Since widespread suffering can be expected to occur if public policy fails to respond
directly to the dire predictions offered by climate scientists and
the optimistic predictions offered by agronomists and other related food scientists, contesting the technical conclusions offered
by these experts entails a lapse in moral judgment—a lapse that
the unnecessarily injured citizens of the future will be justified
in morally condemning. This combination of utilitarianism and
counterfactual narration is not a new mode of thought; it has long
typified dismissive discussions of biotechnology critics.

For example, Michael Fumento, director of the American Security Council Foundation and author of *Bioevolution: How Biotechnology Is Changing Our World* (2003), contends that had India's
public policy been guided by organic advocate Vandana Shiva's
criticisms of the Green Revolution, catastrophic disaster would
have resulted:

If developing world farmers took her one-tenth as seriously as do Western activists and *Time* magazine,
Shiva's proclamations would lead inexorably to massive
famine! Organic farming simply cannot produce the
yields that farming using chemicals or genetically engineered crops can. (Fumento 2002)

Similarly, Ismail Serageldin, director of the Consultative Group
on International Agricultural Research, remarks:

I ask opponents of biotechnology, do you want 2 to 3
million children a year to go blind and 1 million to die

of vitamin A deficiency, just because you object to the way golden rice was created?" (Bailey 2001). Framed in this way, biotechnology critics who fail to ensure that their advocacy is feasible in light of counterfactual exigencies engage in shameful idealism, "where principles prevail over people. (Fumento 2002)

In sum, then, the catastrophe ethics perspective stipulates that publicly differing from the core group of experts on technical matters under dire conditions is tantamount to offering reasons for politicians and the general public to act in an irresponsible manner, one that permits the occurrence of preventable disasters. While such contrarian statements may be legally acceptable—and thus not subject to juridical punishment like hate speech or Holocaust denial—catastrophe ethics proponents believe that the utterances under consideration satisfy rational criteria for determining which speech acts deserve moral censure. In this sense, because catastrophe ethics concerns behavior that falls on the extreme end of the moral spectrum, it is a context in which the liberal democratic priority given to free speech and cognitive autonomy is construed as being trumped by the primacy of survival. catastrophe ethics thus poses a distinctive challenge to theorists and practitioners who are concerned with the fundamental difficulties that expertise poses to public policy because it offers a scenario in which critics may be entitled to use normative terms to censure epistemic claims. If such entitlement exists, then the negative consequences that can follow from critics' creating moral pariahs also may be justified—for example, contrarians may find it difficult to secure funding, have exchanges with experts in the field, acquire tenure, and so on.

In what follows, we discuss catastrophe ethics by focusing on activist judgments instead of ones exercised by average citizens or political leaders. The average Joe or Jane who possess only superficial knowledge of a technical topic should not be held morally accountable for his or her technical assessments, because such a person's views on the matter should not, qua technical view, be taken seriously to begin with. While principles of participatory justice and prima facie political equity endow vulnerable stakeholders with the social right to contribute to discussions about risk, such a right concerns the legitimacy of their vetoing risks that they deem undesirable and inappropriate. However, because veto rights derive from nonepistemic political principles, they have no bearing upon a stakeholder's intellectual capacity to make technical risk assessments.

With respect to political leaders, the relation between politics and expertise frankly seems less contentious. To the extent that politicians in democratic regions are bound by an institutional responsibility to serve the welfare of their citizens, and to the extent that occasions inevitably will emerge where politicians cannot help but act on matters involving technical controversy without the luxury of waiting for additional information to be obtained, we find it understandable that situations will arise where politicians can be rightly condemned on moral grounds for failing to provide democratically acceptable reasons for rejecting policy associated with the consensus view of the core group experts. This is to say that moral condemnation on technical judgment can be appropriate in a political context because the norms of democratic leadership are widely understood to be grounded in a sense of moral responsibility to foster the public good. To this end, when politicians fail to serve the public good as a consequence of rejecting the judgments made by a core group of relevant experts, it would be odd to express our disappointment in the aesthetic terms that we would use to convey a sense of being let down when favored athletes or artists fail to please.

3. The African GM Debate Framed as Catastrophe ethics

For some time it has been alleged that U.S. and European activist opposition to GM food places developing countries—especially in Africa—at risk. This denunciatory narrative has two distinctive features. First, specialist approaches to technological risk assessment are favored over nonspecialist ones. That is, quantitative calculations of expected value are characterized in positive terms, whereas qualitative judgments about safety that are rooted in common lifeworld experiences are demeaned as inappropriate public policy standards. Second, when stakeholders and advocates collaborate with scientists who validate their concerns about GM safety, the following two rhetorical gestures tend to follow. (1) Sometimes, skeptical scientists are depicted as making fallacious appeals to authority and are construed as judging technical matters that fall outside the scope of their expertise. (2) In other cases, a skeptical scientist's expertise is depicted as compromised by ideology.

Relying on these two narrative themes enables GM supporters to frame the debate as a conflict between objective (pro-science) and irrational (anti-science) partisans. What rarely gets acknowledged within such a frame is that genuine scientific disagreement about GM safety still exists, not least because the core community of experts relies on ethical and pragmatic judgments to identify environmental hazards, model exposure, and select populations. This framework also occludes reflexive consideration of historical trends. Many of the rhetorical gestures that typify the GM debate have been endemic to technical controversies; they have long functioned as strategies of legitimation. Additionally, this framework is predicated upon an unduly simplistic conception of the relation between trust and dialogue. We return to this issue in Section 5.

The GM debate is particularly acerbic at present. Erik Stokstad observes that tension between pro- and anti-GM advocates is not diminishing. Commenting on the International Assessment of Agricultural Science and Technology for Development (IAASTD) initiative that began in 2005 with funding from the United Nations, the World Bank, and several countries, he writes:

> On several key issues, consensus proved elusive. Industry scientists and some academics—mainly agricultural economists and plant biologists—believe the assessment was 'hijacked' by participants who oppose genetically modified crops and other common tools of industrial agriculture. Tensions peaked (...) when Monsanto and Syngenta walked out of the assessment. (2008, 1474)

Reflections on the current global food crisis and sustainability are prompting agricultural experts to propose a new Green Revolution (Green Revolution 2.0) that revolves around biotechnology and genetic modification (Montenergo 2008).[5] Of course, there continues to be controversy over the legacy of the first Green

---

[5] As one science reporter put it, given the limited land available for agricultural purposes and the slow pace by which traditional agricultural methods reliably yield abundant crops, this may be the only viable way forward: "The Green Revolution of the 20th century more than doubled the global supply of corn, rice, and wheat. Unless crop yields increase again, however, feeding the earth's 9.2 billion inhabitants in 2050 will require doubling the amount of land under cultivation" (Montenegro 2008). Of course, not everyone views the previous Green Revolution as a success. Consider the following remarks made by MIT anthropologist Stefan Helmreich: "Critiques of the Green Revolution . . . have drawn attention to how the importation of Western agricultural technologies has promoted increased proletarianization among the peasantry; exacerbated inequality by concentrating wealth and credit in the hands of the already rich; aggravated global patterns of dependency on First World capital, expertise, agencies and technologies; wrought havoc on the environ-

Revolution, but this debate must be set aside in the present context. With so many lives at stake, acts that hinder this revolution are being condemned as so misguided as to be immoral. Robert Paarlberg, the Betty Freyhof Johnson Class of 1944 Professor of Political Science at Wellesley College, is playing a key role in this discussion.[6] Stokstad quotes Paarlberg claiming that the IAASTD report lacks scientific credibility and is so broad-based as to be "more a collection of opinions than an incisive summary of the scientific literature" (1476). In his 2008 book *Starved for Science,* Paarlberg emphasizes the following ideas:

1. People are starving throughout Africa. Even though the relevant community of core group scientists finds GM food to be safe, anti-GM activists in Europe and the United States depict it as risky and contribute to decision-making processes that prevent needed hunger relief from taking place, including the acceptance of GM food donations from other nations and legalized opportunities to grow GM food within much of Africa.

2. Given the needless suffering that activist obstructionism causes, nongovernmental agencies (NGOs) that take the elimination of hunger and poverty as their starting points and express a favorable view of GM food (e.g., Oxfam International and Bread for the World) deserve to be characterized as good. By contrast, anti-GM NGOs deserve to be characterized as bad. They export nonscientific and contextually inappropriate conceptions of environmentalism, populism, and organic food purity to Africa (Paarlberg 2008, 99). These deleterious organizations include the International Federation of Organic Agricultural Movements, Greenpeace International, Networking for Ecofarming in Africa, and Food First.[7]

---

ment by introducing new and often ineffective chemical pesticides, fertilizers and high-yielding genetically engineered hybrid seeds (supplied by large multinationals); and jeopardized the livelihood of small farmers who have been encouraged to produce categories and quantities of goods according to the exigencies and schedules of global food markets" (1999, 250–51).

[6] Another important text is Herring 2008, but as it mostly expands on Paarlberg's ideas, our attention will be focused on *Starved for Science.*

[7] Paarlberg writes: "Rather than assisting Africa in the delivery of product-

3. In contrast to other political issues that revolve around matters of technical complexity, African politicians essentially have ignored the core group of scientific experts.[8] Bad faith is a leading cause of this tragic lapse in judgment. African politicians identify more with urban European elites than with rural African farmers, and thus heed the advice of European anti-GM NGOs.[9]

4. For political reasons, pro-science institutions, including the U.N. Environmental Program and the World Bank, have not been strong advocates of GM food. Their lack of advocacy makes them partially accountable for the current state of African disempowerment.[10]

---

ive farming technologies, wealthy nations and NGOs are assisting Africans in keeping such technologies at a distance, likening GMOs [genetically modified organisms] to 'contamination' and regulatory biotechnology to hazardous waste. The GMOs currently on the market have helped farmers become more productive while doing no documented harm to human health or the environment. . . . Instead of helping Africans become more productive and better fed, these donors and NGOs are telling them to become more vigilant: Still poor and hungry, but at least safe from GMOs" (2008, 186).

[8] Paarlberg writes: "Missing in this case is the role sometimes played by transnational networks of scientific and technical experts, given the name 'epistemic communities' by the political scientist Peter Haas. . . . When governmental leaders are uncertain about the implications of a policy choice, expert opinions by scientists can be decisive. . . . However, in the case of agricultural GMOs, the consensus view of scientists has largely been ignored by policy makers in Europe and also by African governments following Europe's lead. The consensus view among scientific authorities is that none of the GM foods or crops commercialized so far has created any new risk to human health or the environment" (2008, 183–84).

[9] Paarlberg writes: "In the end it is not the citizens of Africa who are rejecting agricultural biotechnology. The technology is being kept out of Africa by a careless and distracted political leadership that pays closer attention to urban interests and inducements from outsiders—from European donors, UN technical advisors, NGOs, and export market customers—than to the needs of their own rural poor" (2008, 195).

[10] Paarlberg writes: "The decision of other institutions to duck the controversial GMO issue helped give UNEP [United Nations Environmental Program] more room to operate. The traditionally pro-science organizations that might have been expected to promote GMOs—such as FAO, the World Bank, and CGIAR—mostly held back or backed off, so as to not jeopardize European support and funding" (2008, 189).

5. Anti-GM activists in Europe and the United States have used misleading rhetoric to misrepresent technical issues as cultural ones. For example, when activists insist that by embracing GM crops Africans risk losing their traditional knowledge and skill, the real conditions of material existence become obscured. These conditions are so severe throughout much of Africa that traditional agrarian knowledge, skill, and technology are rendered fairly useless.

6. Although anti-GM activists in Europe and the United States act as proxies for African farmers, most of these farmers disagree with the views that such activists advocate for. This gap between theory and practice is a long-standing trope in subaltern history. It reveals that anti-GM activists substitute misplaced idealism for genuine advocacy.[11]

7. Hypocrisy taints European and U.S. anti-GM activism. Resistance to GM food in rich countries largely arises as a consequence of most citizens in these countries—that is, citizens who are not farmers, seed company employees, or patent holders—lacking clear and direct material benefit from GM purchases. Rather than rallying behind an economic slogan like "We oppose GM food because producers and distributors absorb all of the profit," activists characterize their intentions in terms of moral ideals. The moral claims revolving around the precautionary principle are undermined in practice, when, in medical contexts, reservations about genetic engineering are set aside. Here, the desire for new drugs that can benefit large numbers of first-world people trumps platitudes about risk aversion.[12]

---

[11] Herring writes: "Despite limitations on cultivars and traits, transgenic crops have been accepted by farmers with alacrity, when affordable and available, although access is still limited by politics in many places—some of them the poorest on earth. Farmers have experimented with transgenics, adopted them when their traits have proved useful and have often acquired the technology even at the risk of prosecution. Opposition to transgenics has come not from farmers, by and large, but from those with much less direct interest in agriculture" (2008, 459).

[12] On this issue, proponents of GM food deserve some of the blame. They have misperceived anti-GM advocacy as stemming from scientific ignorance and value-based rejections of genetic engineering instead of lack of reflexivity and egoistic concerns about material benefit.

8. Although early instances of anti-GM activism may be justifiable, contemporary cases are illegitimate. Much of the initial resistance to GM food was fueled by two causes: (1) painful awareness—heightened by the memory of emotionally resonant historical disasters—that some scientific experts and global corporations are untrustworthy, and (2) confrontations with dogmatic advocates who treated all criticism of GM food as irrational. Over time, though, the African food crisis persisted, no reputable scientific study proved that currently available GM food poses a serious health risk, and blanket denunciations of multinational companies like Monsanto being bent on creating "bioserfdom" became increasingly less credible.[13] In light of these outcomes, GM crop skepticism is as out of sync with contemporary trends as blanket associations of science with racism that are anchored in the Tuskegee Syphilis Study that began in 1932.

This eight-point summary clarifies how three issues ultimately drive the moral condemnation of anti-GM activists: *positional vulnerability, self-deception,* and *deprivation.* Positional vulnerability is the charge that when European and American anti-GM activists make their case to Africans, they do not present ideas for consideration that can be freely accepted or rejected. Instead, such activists are seen as co-opting the groups they address by exploiting Africans' lack of agency. Indeed, a conception of Africans as a radically passive group of people runs throughout *Starved for Science.* As suggested above, Paarlberg depicts African citizens as dependent on their political leaders, and African political leaders as culturally mystified and so desperate for NGO donor funds that they cannot reject any potential infusion of money or the conditions that need to be accepted in order for European and American donations to be relinquished.

Ultimately, the question of whether Africans are as vulnerable as Paarlberg depicts them, or whether such a depiction is an artificial and historically deep construct, can be decided only through empirical study. Here, we can only speculate that micro soci-

---

[13] While the pervasive trend of money being privileged over social justice is a genuine problem, dissenters are not acknowledging the value of the new discussions concerning how to best license intellectual property while creating protections for humanitarian use and subsistence agriculture, and how to advance open-source biotechnology projects.

ological and anthropological analyses likely would yield a more
nuanced picture of agency than the one Paarlberg presents. In
the next section, though, we reconstruct the implicit moral argument made against anti-GM advocates in *Starved for Science*. In
so doing, we address the issues of self-deception and deprivation—
issues that we begin to take a stand on in Section 5.

4. The Moral Argument in *Starved For Science*

Although Paarlberg does not appeal to the discourse of moral
philosophy to present his views on the obligation to support to
GM food in Africa, his position can be reconstructed through two
utilitarian premises. In presenting the reconstruction here, we
draw from research conducted by Paul Thompson, which appeals
to Peter Singer's early philosophical work on famine relief as an
anchor point for articulating the first premise.

Singer asserts:

> If it is in our power to prevent something very bad
> from happening, without thereby sacrificing anything
> else morally significant, we ought, morally, to do so.
> (1972, 230)[14]

He maintains that if we can act in such a way as to prevent
a catastrophe from occurring without recourse to morally compromised behavior, we cannot acquit ourselves of the obligation
to counter the expected misery-causing event by appealing to self-
interest (i.e., by emphasizing inconvenience or noting that the
requisite action requires an expenditure of resources) or by observing the pervasiveness of selfish behavior (i.e., by pointing out
how rarely others work to prevent the catastrophe).

Singer often evokes the following example to illustrate this point
(2004, 188–89). If a person can save the life of someone in a developing country by contributing $200 to a reputable NGO like
Oxfam or UNICEF, that person is not confronted with a super-
erogatory situation in which the act of donating is tantamount to
a nice gesture that falls outside the scope of one's basic responsibilities to others. For Singer, failing to make the donation is an
act of neglect, one that is morally identical to allowing a child to

---

[14] Singer is a committed utilitarian, but he contends that this premise is so
basic and obvious that it is deducible from a variety of moral theories and
common sense.

drown in a puddle so as to avoid making a minor sacrifice (i.e., ruining one's clothes and showing up late to work as a consequence of wading in to perform the rescue operation).

In order to extend Singer's claim about justice from financial donation to supporting GM crops in Africa, an additional utilitarian premise is required. The second premise can be found in Norman Borlaug's public speeches and published writing. Borlaug—who co-wrote the preface to Paarlberg's book with former U.S. president and Nobel Peace Prize–winner Jimmy Carter—was an American agronomist who died in 2009. Borlaug's work to increase yields in crops such as maize and rice in order to decrease world famine led to his receiving several prestigious awards, including the Nobel Peace Prize, the Presidential Medal of Freedom, and the Congressional Gold Medal. In a *New York Times* editorial, Steven Pinker (2008) went so far as to suggest that Borlaug may be more admirable than Mother Teresa, because his work on the Green Revolution has been credited "with saving a billion lives."

In their joint preface to *Starved for Science*, Jimmy Carter and Borlaug write:

> In 2000 a joint U.S.-European Union Biotechnology Consultative Forum was appointed by the presidents of the United States and the European Union to look at the full range of issues that have polarized thinking about biotechnology, especially in food and agriculture, on each side of the Atlantic. Although significant differences of opinion existed—mainly related to the regulatory structure involved with certifying agri-biotech products—most of the twenty U.S. and European experts on the panel agreed that agricultural biotechnology holds great promise for dramatic and useful advances in the twenty-first century. In effect, it confirmed the views of the most prestigious national academies of science in North America and Europe (including the Vatican), none of which found any new risks to human health or the environment from any of the applications of crop biotechnology commercialized so far, and all of whom confirm the potential of genetic engineering to improve the quantity, quality, and availability of food supplies. Even so, the debate about the safety and utility of genetically modified (GM) crops continues to grow, enough to discourage governments

in Africa from approving the technology for commercial use. This is a rich-world argument that is hurting the poor. Although there have always been those in society who resist change, the intensity of the attacks against GM crops from some quarters is unprecedented, and in certain cases, even surprising, given the potential environmental benefits that such technology can bring by reducing the use of pesticides. (Paarlberg 2008, viii)

They also claim:

In the past five years, the presidents of several African countries facing widespread drought, crop failure, and hunger have even banned the distribution of donated maize from the United States as food aid, having been told by antibiotechnology groups that this food was 'poison' because it contained genetically modified kernels. Based on such misinformation, they have been willing to risk thousands of additional starvation deaths rather than distribute the same maize that regulators approve in the United States and that Americans have been eating for more than a decade with no documented ill effect. (viii–ix)

Thompson defines the "Borlaug hypothesis" as follows:

If agricultural biotechnology and the development of GM crops have the potential to contribute to a lessening of hunger and deprivation over the long term, then people have a moral obligation to support the use of these techniques, at least in so far as they are deployed in pursuit of that end. Furthermore, this obligation overrides less compelling ethical concerns that may exist for GM crops (2007, 219).

Put in the language we used in section 3, the Borlaug hypothesis concerns the moral obligation not to *deprive* people of tools that they can use to meet basic needs, such as having access to enough food to live. To avoid potential misunderstanding of the Borlaug hypothesis, two caveats are in order. First, the proviso about "overriding less compelling ethical concerns" does not imply that all ethical reservations about GM crops are misguided. Rather, when this proviso is understood in the broader context of Borlaug's extensive comments on GM crops, it can be understood as a claim about what Thompson calls "side constraints." Thompson notes fairly extensive conditions that would need to be met in order for *any* agricultural development project to be considered ethically justified. Although conditions such as "respecting the agency of recipient peoples" and "carefully weighing environmental risks" might be inherently vague or even contentious in themselves, there is nonetheless a discernible sense in which they have nothing to do with the use of biotechnology as such. For Borlaug, concerns that GM crops will provide economic benefit to global elites at the expense of local farmers, and concerns that indigenous values are threatened by the prospect of integrating GM crops into local practices, are best conceived of as reservations that particular forms of public policy will promote injurious outcomes. Crucially, they are not rejections of access to GM crops in all possible circumstances. Nor are they claims about GM crops being inherently immoral. As Borlaug sees it, social justice advocates who are committed to combating first-world exploitation of Africans are people who are necessarily committed to finding appropriate strategies of GM implementation, dissemination, and property rights. Given this way of framing the GM controversy in Africa, Borlaug implicitly holds activists who condemn GM crops *tout court* morally responsible for *deceiving themselves* in believing that that there are legitimate issues to debate beyond side constraints. (It is, perhaps, important to make a clarification not germane to the central argument at this juncture, though vital for clarifying authorial intent. The discussion of self-deception is Selinger's interpretation of Borlaug. Thompson interprets Shiva-style objections as empirical claims to the effect that side constraints have not been met. From his perspective, it is the failure to *demonstrate* that side constraints have indeed been met—if they have been—that is material to the failure of the Borlaug hypothesis on ethical grounds.)

But such comprehensive and total objections to GM crops do not exhaust the debate. As noted, side constraints are important because for philosophers like Thompson the legitimacy of GM depends upon doing so only when side constraints have been met. Borlaug and Paarlberg are not entirely clear on this point. It is possible that they would argue either that these constraints are in fact being met (at least to a reasonable degree), or that they would acknowledge room for improvement over past Green Revolution initiatives. But their text also leaves open the possibility that they think the need to enhance productivity is so compelling as to effectively vitiate all side constraints. As Thompson interprets the objections to GMOs made by opponents, they are claims that side constraints have not been met. It is thus the failure to demonstrate that side constraints have indeed been met that is material to the failure of the Borlaug hypothesis.

The second caveat also concerns self-deception. If the Borlaug hypothesis applied only to people who are confident that GM crops can increase crop yield significantly in comparatively shortened periods of time without posing a significant threat to humans or the environment, it would have little normative significance. Perhaps only scientifically knowledgeable racists—for example, people who wanted to deny Africans a useful technology because they hate non-Caucasians—would fall within the extremely narrow scope of such an interpretation. Instead, Borlaug holds activists accountable for not supporting agricultural biotechnology in Africa because he believes there is overwhelmingly strong scientific evidence supporting the safety and utility of GM food. In this sense, while Borlaug can be sufficiently fallibilist to allow for the possibility that he and the core scientists he agrees with are wrong about risk, he still can insist on the following: Given the likelihood that the core group is right, anti-GM advocates have a responsibility to approach the hunger crisis with viable solutions. The core group provides a basis for positive action in response to a morally compelling need, while opponents who question aspects of the core group have offered no alternative plans to address this need. Given the weight of the evidence favoring the core group, this failure is itself a morally culpable act. The anti-GM activists have deceived themselves into thinking they are working to end hunger, while from Borlaug's perspective they have no concrete plan at all.

It is, however, important to bound this qualification. While the above seems to be a reasonable summary of the position to which Borlaug and now Paarlberg are committed, the factual question of whether anti-GM advocates *have* or *have not* offered alternative plans of action cannot be addressed in the present context. Many GM opponents certainly see themselves as having identified alternatives in the form of fair trade initiatives and sophisticated methods that involve alternative crops, infrastructure development, and land reform efforts. The point is relevant to the extent that one sees Borlaug and Paarlberg as acquiring an obligation to discuss and rebut these claims of opponents when they assert, on the basis of their expertise, that GM crops are the most viable and only morally responsible alternative.

5. Deflationary Position: Appealing to Discourse Ethics and Virtue Ethics

Thompson's research also provides the basis for the deflationary philosophical counter to the Borlaug hypothesis. Thompson insists that the counter does "not derive from facts about GMOs or their fate in the environment" (2007, 225). Instead, it focuses on character as an indicator of trustworthiness and respectful conversation as a morally assessable discursive form that, *pace* norms of discourse ethics, obligates interlocutors to exchange ideas charitably.

Drawing from extensive personal experience, Thompson finds that a significant portion of the experts representing the relevant scientific disciplines in the GM debates have, as a matter of course, failed to execute the responsibilities of expertise. By exhibiting a "tendency to misrepresent" stakeholder objections as baseless "obstacles" that should be "set aside through whatever means" are available, they have relied solely on the authority they possess in virtue of their credentials, and have actually denied the public any opportunity to consider the case in light of evidence and argumentation (2007, 224). In making patronizing assumptions about the public's knowledge, abilities, concerns, and capacities to generate meaning, these experts have used the tools of instrumental reason to displace serious moral thinking with strategic or manipulative argumentations and thereby rushed to disqualify public views from having any authority in debates about risk assessment. These tactics are so disrespectful that they legitimate the public making the following inductively valid virtue-theoretic argument: Since these GM advocates conduct themselves in an untrustworthy manner, they are likely to espouse untrustworthy conclusions about GM safety. Put in discourse ethics terms, since these pro-GM scientists who monopolize the public sphere conceptualize the public as so irrational as to be in principle incapable of making meaningful contributions to debates about GM safety, they fail to adhere to the morally required principle of charity. Thompson thus concludes that the Borlaug hypothesis fails:

> My personal, anecdotal assessment is that the truly virtuous are roughly offset by the truly disreputable, leaving the field to the dismissive and busy. This tips the balance towards the less favorable assessment of in-

sider's virtue when they are viewed as a group. Even
among those who take up the pen to write in favor of
biotechnology, very few display any evidence of having
read any of their opponents' views. Citations to opponents' views are even rarer, and patient attempts to
restate and fairly represent opponent's positions before
launching into the pro-biotech agenda are the rarest
of all. As such, I conclude that if there is blame to
be distributed for the hostility that even benevolent
applications of agricultural biotechnology now face, a
large share of that blame must be shouldered by the
agricultural community itself. (2007, 231)[15]

To ensure that Thompson's position on the GM food debate
is properly understood, it will be useful to extend his view to
a hypothetical case where the American public refuses to accept
scientific conclusions about risk and GM food, even though at an
institutional level the scientists who convey the risk analysis have
exhibited all of the virtues requires by discourse ethics. For this
purpose, a thought experiment will do. Imagine that twin earth is
identical to our earth in all respects except that it has a different
history of GM food. On twin earth, there is a grassroots push in
the United States to make the production of GM foods illegal, even
though scientists have always discussed GM food with the American public in ways that fully accord with the norms of discourse
ethics. Let's further imagine that on twin earth the only concerns
expressed by American citizens and activist groups about GM
foods are ones concerning potential health risks, even though GM
foods are no riskier on twin earth than they are on our earth. In
this case, the view that Thompson endorses would place responsibility on the government of twin earth to not to bow to public
pressure. Furthermore, if we presume that on both twin earth and
regular earth the Borlaug assertion that many people will die as
result of rejecting GM crops is both true and well supported by

---

[15] To avoid potential misunderstanding about his conclusion as to why the
pro-GM advocates have—at least for now—lost the right to be viewed as
credible, Thompson adds: "Lest any stray outsider be confused by my argument, this conclusion is *not* an anti-biotechnology conclusion. Recombinant
techniques for developing new crops *should be* deployed in the fight against
hunger, and ordinary citizens not only *should* support this deployment, but
should seek ways to ensure that the industrial world's taste for non-GM crops
does not preclude the use of biotechnology to help the hungry" (2007, 232).

scientific evidence, then the Thompson view would justify finding twin earth activistswho are advising developing countries to ban GM food guilty of immoral and potentially genocidal conduct. The main difference, then, between Thompson offering such a critique of activists on twin earth but criticizing Paarlberg's reliance on the Borlaug hypothesis on regular earth is that scientists on regular earth have not acted according to the norms of discourse ethics. From Thompson's perspective, Paarlberg puts premature blame on those members of the public who have been persuaded by anti-GM activists.

If the conditions of discourse ethics ever were met, then Thompson's position would require anti-GM activists who dissent from the core group of scientific experts to be morally obligated to frame their case to the public with a robust set of qualifiers that effectively undermines the persuasiveness of their case.

6. The Sociological Rejoinder

The program known as the third wave of science studies would approach the matter in a different way (Collins and Evans 2002, 2007), one that seems to imply a change in the locus of moral responsibility. The third wave is an attempt to find a way of making real-time assessments of scientific worth that still take into account the revolution in our understanding of scientific knowledge that has occurred since the middle of the twentieth century. While the so-called first wave of science studies tried to explain and find ways to nurture science's unquestioned authority in matters technical, the second wave of science studies deconstructed science.[16] It showed that when science was examined very closely it was not so different from ordinary life and ordinary reasoning.[17] What

---

[16] For the first wave see the works of philosophers of science, including Hempel 1966 and Popper 1992, and the works of sociologists, notably Merton 1973.

[17] The second wave was essentially a child of the 1960s. For the second wave see, for example, Bloor 1973 and 1976; Collins 1975 and 1985; Collins and Pinch 1993 and 1998; Latour and Woolgar 1979; and many publications in the journal *Social Studies of Science*. The second wave was the target in the so-called science wars. It is only fair to add that the three-wave metaphor has been resisted by many leading figures in the enterprise known as social studies of science and technology. The metaphor is said to gloss many fine distinctions—which it certainly does—and, of course, it is never nice to have one's personal discoveries described as merely part of a broad historical movement.

this means is that the outcomes of scientific controversies are co-extensive with the outcome of essentially "political" arguments about which scientists and which results are to be believed and trusted. Scientific history, under the new model, was continuous with political history, and the only way to understand why one result emerged from scientific debate rather than another was a retrospective analysis of the particular passage of history. Underpinning this approach was the demonstration that scientific evidence was available to many interpretations—there was a high degree of "interpretative flexibility"—meaning that different scientists, given the same evidence, could reasonably argue for either p or *not-p*, respectively, for an indefinite period. Typically, decades would pass before all dissenters to a consensus position disappeared: as Max Planck is said to have put it, science advances funeral by funeral.

The third wave could not be a return to the first wave. The skeptical positions developed from the 1960s onward could not just be forgotten, and the results of the detailed studies of scientific debate conducted from the early 1970s onward, which showed the mundane face of science and the fact that there could be a coherent logic to each the competing position in a contested area, were not going to go away. Instead the third wave put forward studies of expertise and experience (SEE) as the solution to the problem of technical policy making in real time. It would not be a solution anything like as ambitious as that canvassed by the first wave. SEE could not deliver the truth and would frequently lead to wrong choices; nevertheless, it remained the only way to go. Quite simply, it claimed that the only reasonable thing for a policy maker to do was to prefer the technical advice of those who "knew what they were talking about" rather than those who did not. SEE was a matter of working out what it meant to "know what you were talking about" and who knew it, by classifying kinds of expertise. Expertise could be gained through joint experience in a technical domain, particularly scientifically informed experience, and/or by absorbing the "tacit knowledge" of the domain experts. The solution for the policy maker was then to choose between competing technical choices on offer by choosing between the experts on offer, always preferring those with the soundest expertise and most worthwhile experience. To repeat, this did not guarantee that the right technical choice would be made, but it was the least bad way of doing things. Of course, making a technical choice is not making a political decision, and it might still be that political values could override a technical opinion, especially where the latter was heavily disputed. What was vital, however, was not to disempower voters by pretending that a political choice was a scientific choice where the science that informed it was the position of a clear minority, or a group of mavericks or outcasts.

The difference between this position and that of Thompson seems to lie in the duty of the minority group—the mavericks and outcasts. Under wave three, it is the responsibility of politicians and public opinion formers, such as journalists, to weigh the contributions of the different kinds of experts and expertises on offer so as to make a properly balanced scientific choice.

In the case of journalists, this notion of "balance" is very different to the notion of balance in the expression "balanced story." For example, in the United Kingdom there has been a controversy over the efficacy and danger of the mumps, measles, and rubella (MMR) vaccine. A single medical doctor, Andrew Wakefield, claimed that the vaccine could cause autism, while the overwhelming majority of the medical profession, including epidemiologists, argued that there was no connection. The duty of journalists here should have been to weigh the expertise and conclude that Wakefield's views were not yet worth reporting because of the lack of balance with consensual medical and epidemiological opinion. Instead, they tried to write what they thought of as "balanced stories" in which medical and epidemiological opinion was given even weight with accounts of parents whose children had been diagnosed with autism after having an MMR jab. (It is, of course, statistically inevitable that there will be a number of such cases even if MMR is completely safe.) In the case of policy makers something similar applies: the policy makers should be able to weigh the vast and almost complete consensus on the one hand against the single maverick view on the other and make the reasonable decision to go with the majority.

Under wave three the same institutional responsibility does not fall on the maverick scientists. So long as they are not engaging in deceptive behavior, mavericks should be free to be as vigorous in their advocacy of their views as they like, even when, in an ideal world, the norms of discourse ethics have been satisfied, such as the hypothetical case of twin earth discussed in section 5. That is because wave three does not reject wave two, and wave two demonstrates that a maverick view can be rationally held in science. Indeed, science as we love and admire it is almost defined by the possibility of holding a maverick view. Galileo was a scientific hero because he stood up to the might of contrary opinion. Who could more perfectly fulfill the job of role model for the aspiring scientist, or even the ordinary citizen, than the scientist who holds a minority view all his life and is proved right only after death? Under both wave two and wave three, therefore, it is the positive duty of mavericks to stick to their sincerely held views and even try to proselytize, while it is the equal duty of policy makers to ignore them once they have been given a reasonable run for their money, and it should be the duty of the members of the third estate to mention them in passing if necessary but not to boost their credibility by treating their claims with symmetrically with near-consensual scientific opinion.

One might say that this position arises from privileging "actors' categories." The first duty of the sociologist of scientific knowledge is to try as hard as possible to enter the world of the actors and to understand things from their perspective. If one does this properly, and if one's scientist actors have scientific integrity and are doing their best to act as scientists, not politicians or whatever (something assumed throughout this exposition of the third wave), one discovers that scientific mavericks see themselves not as mavericks but as the only people with a proper grip on the truth. Equally truly, the members of the core group representing the growing consensus see the mavericks as fools, or people acting out of inauthentic motives, or trouble makers, or something like that— and they hold their view sincerely too. That is just how scientific argument goes. To turn to Collins's (2004) long study of gravitational waves, the late Joseph Weber claimed as the 1960s became the 1970s to have detected the waves. By 1975 no one believed Weber, but against all opposition, and in the face of accusations of foolishness and charlatanism, he stuck to his claim until he died (in 2000). In 1996, another gravitational wave physicist, part of what by then had become a billion dollar international enterprise sparked by Weber's initial efforts, put it this way:

> If Weber had done the smart thing for him—politically, socially, and everything else—he would have—but maybe not the true, not really the correct thing to have done given his beliefs—would have been to say, Well, yes, I admit I was wrong, and all this,' and then he would be the venerable leader of the field. And, er, we would all acknowledge 'Uncle Joe' in every lecture. And people have tried and tried and tried to get him to accept

that role. (qtd. Collins 2004, 811)

The point lies in the qualification "but maybe not the true, not really the correct thing to have done given his beliefs." If we are going to preserve the idea of science as an enterprise capable of giving moral leadership (Collins 2009), then we have to see Weber as a hero for choosing science over adulation and we have to pass the responsibility for choosing between the views of the heroes to those whose job is to make such judgments—journalists and, especially, politicians.

As the preceding passage intimates, the third wave must have not only a normative but also a moral agenda at its heart. It is wave two that makes what was the instrumental choice of science under wave one become a moral choice. Wave one implied that one could get "ought" from "is"—the "ought" of democracy— the political system that, according to the Mertonian norms, best nurtures science—from the "is" of science's efficiency under democracy. But wave two showed that argument did not work—there was no "is" when the process of science was examined closely enough. But wave three still chooses science because it is best— not best for a reason, just plain best. Try imagining a society without any scientific values, and why it is best becomes self-evident. This choice is called "elective modernism" (Collins 2009). The third wave says, "Choose scientific expertise above other expertises." That choice requires, of course, that scientific expertise first be demarcated from other expertises.[18] But policy makers are then required to choose which body of technical experts to take most seriously, and the moral obligation upon them is to try to do it by reference to the weighing of expertises already discussed and, especially, not to pretend that the existence of a pocket of dissensus licenses any political decisions dressed up as scientific doubt. To hide political decisions under a cloak of faux scientific disagreement is to take real choice away from voters and disempower the political process.[19] There is always doubt in science— that is its strength as a system of knowledge making and a moral beacon, but that does not mean scientific understanding belongs

---

[18] An old problem that is addressed again in chapter 5 of Collins and Evans 2007 but certainly needs more analysis.

[19] As seem to have happened in South Africa with Thabo Mbeki's refusal to allow the state to sponsor antiretroviral drug treatment of pregnant mothers with HIV (Weinel 2007).

to everyone.

Conclusion

The following summary of sections 5 and 6 distills the contrasting disciplinary perspectives on morality and activism.

- Readers who follow Thompson's philosophical lead must conclude that once the ideal conditions of discourse ethics have been satisfied, the minority of dissenting experts are morally obligated, even under conditions of crisis, to present such charitable and elaborate reconstructions of how the core group has, in good faith and with good reasons, rejected their position, and that they convey their views in the public sphere through humble speech and gesture. The minority of dissenting experts thus should proceed—to use virtue ethics language—"modestly," constantly reinforcing that their judgments represent an outlier position that the core group has rejected after due consideration. Moral responsibility thus is attributed to experts at the level of individual behavior. On a practical level, it can be expected that by proceeding modestly dissenters will effectively undermine the persuasiveness of their claims.

- Readers who endorse the sociological perspective of the third wave must conclude that moral responsibility for dealing with dissenting scientific claims in the public sphere belongs solely to political leaders and members of the press. Even when the ideal conditions of discourse ethics have been met, the third wave grants primacy to a conception of intellectual freedom that is rooted in the epistemic reality of interpretative flexibility. It grants nondeceptive dissenters moral license to aggressively push their technical views on everyone—politicians, the press, and the public—as hard and heroically as they please.

In the final analysis, a major difference exists in the type of affect, demeanor, and rhetoric that the philosophical and sociological views find morally permissible for dissenting experts to display in the public sphere once the members of the core group of experts have virtuously satisfied the ideal conditions of discourse ethics.

To concretize this point in terms of the debate about African political leaders being advised to ban GM food, the following conclusion—already intimated in the twin earth discussion—can be drawn.

- For the third wave proponent, nondeceptive, technically expert, activists should not be, under any circumstances—*including conditions where the ideal norms of discourse ethics have been met*—considered immoral if, on technical grounds, they aggressively counsel African leaders to ban GM food. The job of assigning praise and blame falls to social and political actors. By contrast, from the philosophical perspective that Thompson endorses, aggressively pushed anti-GM counsel articulated on technical grounds through the rhetoric of misunderstood scientific heroism qualifies as nonvirtuous behavior *if it occurs after the core group has satisfied the ideal norms of discourse ethics.*

Since these views only diverge at the limit (when ideal conditions have been satisfied), the disciplinary differences do not bear upon current empirical reality in a neat way. As discussed in section 5, significant portions of the core group of pro-GM experts have not engaged in virtuous conduct, and for Thompson this releases the anti-GM activists from their obligations to qualify their claims. It does not change the third wave position, but it remains a constant of that position too that scientific argument should be conducted with integrity, which is probably not the case in the current situation. And, to be fair, it would seem that some of the anti-GM activists have engaged in comparably uncharitable behavior. Nevertheless, there is much to learn from considering idealized cases. The material presented here has, one hopes, shed light on the oversimplifications underlying Krugman's conception of "catastrophe ethics" and clarified the difficulty of holding activists morally accountable for their technical council. One hopes too that it has also presented a set of clear conditions under which it becomes possible to use clear and principled differences to debate the moral status of dissenting activists.

Bibliography

Bailey, Ronald. 2001. "Dr. Strangelunch, Or: Why We Should Learn to Stop Worrying and Love Genetically Modified Food." *Reason Online.* (January.)

Bloor, David. 1973. "Wittgenstein and Mannheim on the Sociology of Mathematics." *Studies in the History and Philosophy of Science* 4:173–91.

———. 1976. *Knowledge and the Social Imaginary.* London: Routledge and Kegan Paul.

Caroll, Rory. 2002. "Zambia Slams the Door Shut on GM food." *The Guardian* (online). (October 30.)

Collins, Harry. 2009. "We Cannot Live by Scepticism Alone." *Nature* 458 (March): 30–31.

———. 2004. *Gravity's Shadow: The Search for Gravitational Waves*. Chicago: University of Chicago Press.

Collins, Harry, and Robert Evans. 2002. "The Third Wave of Science Studies: Studies of Expertise and Experience." *Social Studies of Science* 32, no. 2: 235–96.

———. 2007. *Rethinking Expertise*. Chicago: University of Chicago Press.

Collins, H. M. 1975. "The Seven Sexes: A Study in the Sociology of a Phenomenon, or The Replication of Experiments in Physics." *Sociology* 9, no. 2:205–24.

Collins, H. M. 1985. *Changing Order: Replication and Induction in Scientific Practice*. Beverly Hills, Calif.: Sage. (Second edition Chicago: University of Chicago Press, 1992.)

Collins, H. M., and Trevor Pinch. 1993 and 1998. *The Golem: What Everyone Should Know About Science*. Cambridge: Cambridge University Press.

Fumento, Michael. 2002. "The Villainous Vandana Shiva" *National Review Online*. (August 27.)

———. 2003. *Bioevolution: How Biotechnology Is Changing the World*. San Francisco: Encounter Books.

Hempel, Carl. 1966. *Philosophy of Natural Science*. Englewood Cliffs, N.J.: Prentice Hall.

Herring, Robert. 2008. "Opposition to Transgenic Technologies: Ideologies, Interests, and Collective Action Frames." *Nature Reviews Genetics* 9 (June): 458–63. *Independent, The* [U.K.]. 2008. "Leading Article: There Is No Reason for Blanket Ban on GM Crops." (Online.) (June 20.)

Krugman, Paul. 2008. "Can This Planet Be Saved?" *New York Times* (online). (August 1.)

Latour, Bruno, and Steve Woolgar. 1979. *Laboratory Life: The Construction of Scientific Facts*. Beverly Hills, Calif.: Sage.

Makinede, Martin. 2004. "Agricultural Biotechnology in African Countries." In *Cross-Cultural Biotechnology*, edited by Michael Brannigan, 115–26. Lanham, Md.: Rowman and Littlefield.

Merton, Robert. 1973. *The Sociology of Science: Theoretical and Empirical Investigations*. Chicago: University of Chicago Press.

Montenegro, Maywa. 2008. "Green Revolution 2.0" *SEED Magazine* (online). (August 20.)

Paarlberg, Robert. 2008. *Starved for Science: How Biotechnology Is Being Kept out of Africa.* Cambridge, Mass.: Harvard University Press.

Pinker, Steven. 2008. "The Moral Instinct." *New York Times* (online). (January 13.)

Popper, Karl. 1992. *The Logic of Scientific Discovery.* London: Routledge.

Scapp, Ron, and Brian Seitz, eds. 2006. *Etiquette: Reflections on Contemporary Comportment.* Albany: State University of New York Press.

Shapin, Steven. 1995. *A Social History of Truth: Civility and Science in 17th Century England.* Chicago: University of Chicago Press.

Singer, Peter. 1972. "Famine, Affluence, and Morality." *Philosophy and Public Affairs* 1:229–48.

———. 2004. *One World: The Ethics of Globalization.* New Haven: Yale University Press.

Shapin, Steven. (1995). *A Social History of Truth: Civility and Science in 17th Century England.* Chicago: University of Chicago Press.

Stokstad, Erik. 2008. "Agriculture: Dueling Visions for a Hungry World." *Science* 14 vol. 319, no. 5869 (March): 1474–76.

Thompson, Paul. 1999. "The Ethics of Truth Telling and the Problem of Risk" *Science and Engineering Ethics* 5:489–510.

———. 2007. "Ethics, Hunger, and the Case for Genetically Modified (GM) Crops." In *Ethics, Hunger, and Globalization: In Search of Appropriate Policies*, edited by Per Pinstrum– Anderson and Peter Sandoe, 215–35. Dordrecht: Springer.

Weinel, Martin. 2007. "Primary Source Knowledge and Technical Decision-Making: Mbeki and the AZT Debate." *Studies in History and Philosophy of Science* 38, no. 4:748–60.

Wynne, Brian. 2006. "May the Sheep Safely Graze?" In *Risk, Environment, and Modernity*, edited by Scott Lash, Bronislaw Szerszynski, and Brian Wynne, 44–83. Thousand Oaks, Calif.: Sage.

7

Competence and Trust in Choice Architecture

Written with Kyle Powys Whyte

1. Introduction

Richard Thaler and Cass Sunstein's *Nudge: Improving Decisions about Health, Wealth, and Happiness* advances a theory of how designers can improve decision-making in various situations where people have to make choices, or where failing to make deliberate choices results in their behavior being influenced by default settings (See also Thaler and Sunstein 2003 a, b). The following four claims summarize some of the main ideas in *Nudge*:

1. Behavioral economics suggests that people tend to rely on *biases* like mental shortcuts, inclinations, models, gut feelings, and heuristics when they make decisions in situations that do not afford them sufficient time or information, or in situations when people find themselves subject to unanticipated arousal and temptation.

2. It is reasonable to assume that reliance on biases influences people to make some ill-informed and costly decisions about their health, financial security, and pursuit of the good life. The costs are often borne by other members of society. Examples of bad decisions range from choosing the wrong health or retirement plans when interfacing with the forms, paper or electronic, that human resource departments give to new employees, to driving too fast through a treacherous curve while taking in a scenic stretch of highway.

3. *Choice architects* design those aspects of technologies, interfaces, and built environments that present users with distinctive opportunities (which we will refer to as the choice context) for interacting with people, objects, and surroundings. Their work is ubiquitous, present in computer interfaces, credit card consoles, strategically arranged merchandise, strongly worded contracts, and so on. They can help

to improve choices and behavior by subtly calibrating the
choice context to work with peoples' predictable tendencies
to rely on biases. These calibrations are called *nudges*

4. It is reasonable to expect that nudges will, on average, cut
the costs of bad decisions and thereby increase savings to
individuals and, in some cases, other members of society.

We argue in this chapter that the moral acceptability of nudges
hinges in part on whether Thaler and Sunstein can provide an
account of the *competence* required to offer nudges—an account
that would serve to warrant our general *trust* in choice architects.
Our case will be presented as follows. In Sections 1 and 2, we expand upon our initial description of nudges and clarify why Thaler
and Sunstein appeal to the principles of *libertarian paternalism* to
set ethical limits on possible exploitative uses of nudges. In Section 3, we defend the idea that even if libertarian paternalism
is as attractive as Thaler and Sunstein contend, it does not establish whether choice architects can be trusted to offer nudges
that improve our health, wealth, and well- being. What needs
to be considered, on a methodological level, is whether Thaler
and Sunstein have clarified the competence required for choice
architects to subtly prompt our behavior toward making choices
that are in our best interest from our own perspectives. Competence is among the widely accepted characteristics that justifies
the trustworthiness of the testimony and products of scientists,
engineers, technicians, and others who we depend on to influence
and improve our choices and behavior. Choice architects who offer nudges play an analogous role, which suggests that Thaler and
Sunstein should have an account of competence that warrants our
general trust in choice architects, especially since we will not be
aware of the fact we are being nudged most of the time.[1]

But because *Nudge* lacks an account of competence, we want to
identify the salient features that a prospective account would have
to include. In Section 4, we argue that, among other features, an
account of the competence required to offer nudges would have to
clarify why it is reasonable to expect that choice architects can

---

[1] Our treatment of trust is limited to how trust relates to competence.
Based on this treatment, we do not consider whether trust exists in certain
technological interfaces (Taddeo 2009), nor do we address the philosophical
debates on what trust is in social relations (Wright 2009; Baier 1986; Holton
1994; Jones 1996; Hinchman 2005a,b; Hieronymi 2008).

understand the constraints imposed by *semantic variance*. Semantic variance refers to the diverse perceptions of meaning, tied to differences in identity and context that influence how users interpret nudges.[2] We conclude by suggesting that choice architects can grasp semantic variance if Thaler and Sunstein's approach to design is compatible with insights about meaning expressed in science and technology studies and the philosophy of technology.

2. Nudges

In this section, we restrict our discussion of nudges to those aspects which are relevant to our main argument, beginning by noting that the theory of choice architecture and nudges is rooted in an understanding of biases that people are subject to in various situations where they have choices to make (for another description, see Lobel and Amir 2009). The biases that Thaler and Sunstein focus on are ones that affect the quality of our choices and behavior across the spectrum of racial, sexual, and educational differences. Some, like Thaler, classify them as basic constituents of human nature.[3]

Thaler and Sunstein build their account of biases from the basic tenets of *dual-process theory*, a view that stipulates that people's thought is structured by two systems, *automatic* and *reflective* (Epstein 1994; Thaler and Sunstein 2008). Automatic thinking is characterized as uncontrolled, effortless, associative, fast, unconscious, and unskilled (Thaler and Sunstein 2008, 20). It is personified by the *gut reactions* of Homer Simpson, an impulsive cartoon character, and contrasted with the idealized assumptions

---

[2] Readers familiar with Don Ihde's phenomenological philosophy will note that semantic variance bears conceptual affinity with *multi-stability* (Ihde 1977; Selinger 2006; Ihde 2007). We do not use the latter term because it is broader in scope than the former. We selected a more delimited concept for the simple reason that the broad scope of multi-stability has prompted skeptical debate that should not be applicable here (Cerbone 2009).

[3] The following is an excerpt from the transcript of Thaler's interview on the Tavis Smiley's PBS show: "**Tavis**: The research indicates that this varies from men to women, from racial group to racial group, or is this across the board? **Thaler**: Well, there are small differences. I think women pay a little more attention to detail than men. There are other cultural differences, but by and large, humans are all hardwired the same way. We're all busy. We all have trouble controlling our impulses and the kinds of things that we talk about in the book are the universals. **Tavis**: Even for those of us who are more educated? **Thaler**: Absolutely. We're all human. You know, one of the most powerful biases we have is over-confidence."

about reasoning associated with *homo economicus* (22). By contrast, reflective thinking is controlled, effortful, deductive, slow, self-aware, and rule-following (20). Reflective thinking is embodied by science fiction character Mr. Spock's rational, deliberate, and unemotional approach to problem solving (22).

Without reflective thinking, humans could not make careful and effective long-term plans. But since reflective thinking is time consuming and requires people to effectively process good information, it cannot be relied on in a variety of contexts, which not only include situations where we have to act quickly, but also situations where our choices have delayed effects, are difficult in nature, occur infrequently, yield poor feedback, and present choice context that we have little to no experience dealing with. In addition, reflective thinking does not help us in situations where greater triggers for arousal and temptations exist than we initially anticipated (23). In these instances, we turn to the automatic system of thinking, rapidly drawing from prior, often unrelated, experiences and allowing our behavior to be guided by rules of thumb or mental shortcuts developed in those experiences (23).[4] These rules of thumb serve as biases when the experiences from which they are drawn or the evidence to which they appeal are irrelevant to the decision at hand.

An illustrative example of a bias is the use of inappropriate anchors, which are mental shortcuts that compensate for lack of information. Consider what happens when someone is asked to estimate the total population of Milwaukee, but does not know the answer and has to respond immediately. In this case, if the person is from Green Bay, he or she likely realizes that Milwaukee has more people than Green Bay (100,0000 people). As a result, the person could offer the guess that Milwaukee has about 300,000 people. But, if the person is from Chicago (3,000,000 people), then, knowing that Milwaukee is definitely smaller, he or she might guess that the population is 1,000,000 people. Unfortunately, neither answer is very accurate; the actual population is about 580,000. The reason why each person decides on a wrong answer is that he or she uses the city where he or she lives as the basis, or anchor, for making a decision and subsequently makes inappropriate mathematical adjustments. Of course, it is irrelevant

---

[4] In such situations, automatic thinking can provide us with a sensible orientation to what some phenomenologists call practical intelligence and practical coping.

that one person is from Green Bay and the other from Chicago, as the population of neither city has any objective bearing on estimating the population of Milwaukee (Thaler and Sunstein 2008, 23).

Reliance on biases can be costly in situations where we interact with technologies, artifacts, and built environments, and there is often more at stake than correctly answering a trivia question. An apt example is the power exerted by default options. Thaler and Sunstein claim that many US citizens lack the willpower or relevant information about investment strategies to implement a sound savings plan. Motivated by the ease of sticking to the default *opt in* setting found in many employee benefit forms, they fail to save enough money for retirement. This adverse outcome causes personal discomfort and stresses the social security system. A similar example concerns the limited influence exerted by street signs that convey the importance of slowing down for a treacherous curve. Drivers who are already comfortable speeding or who do not have the time or sufficient information to process the upcoming curve can be tempted to stay in their comfort zones rather than heeding the warning to reduce speed. Such inertia can be costly, especially if accidents result. Other biases include but are not limited to the optimism bias (i.e., people have an unrealistic grasp of how good their abilities are), loss aversion bias (i.e., people prefer gains to losses, even in situations incurring losses is in their best interest), confirmation bias (i.e., people have a tendency to overestimate information that reinforces things we already believe), hyperbolic discounting (i.e., people have stronger preference for immediate payoffs relative to later payoffs), focusing effect (i.e., people have a tendency to place too much emphasis on one variable when making predictions about future outcomes), and impact bias (i.e., people have a tendency to overestimate the length and intensity of future feeling states).

Nudges are simple solutions to our problems with biases. Choice architects design nudges when they subtly calibrate how choices are presented to us in order to work with our predictable tendencies to rely on biases. In this sense, a nudge, for Thaler and Sunstein, is "any aspect of the choice architecture that alters people's behavior in a predictable way without forbidding any options or significantly changing their economic incentives" (6). Paying students to study thus does not count as nudging them to be studious.

Such a bribe changes the financial incentive.[5] Additionally, Thaler and Sunstein stipulate that nudges must be "cheap and easy to avoid" (6). Mandates, such as banning junk food or requiring merchants to write their instructions, product labels, warranties, and the like, are not nudges either (6).

Put hyperbolically, *Nudge* is a kind of manifesto that claims that choice architects in all facets of life should get into the business of accounting for biases through nudges. As we read Thaler and Sunstein, nudges do better than educational programs for at least the following two reasons. First, the automatic system is an inherent part of how we think, and even serves us well in some situations. Consequently, it does not make sense to treat automatic thinking as undisciplined reasoning that should be eliminated. Second, since we rely on automatic thinking ubiquitously, no educational program could be comprehensive enough to anticipate every occasion in which we rely on biases, or offer effective rules for practical coping that correlate with all such occasions. In light of these and related reasons, Thaler and Sunstein repeatedly remind readers of *Nudge* that they remain as vulnerable to the sway of biases as everyone else.[6]

Beyond differing from educational programs and proposals that alter behavior through new financial incentives and mandates, choice architecture also can be contrasted with critical humanities and social science theories of design. Notably, philosophers of technology and science and technology studies theorists have

---

[5] See, for example, *Freakonomics* (Levitt and Dubner 2005), for examples of microeconomic fixes. Of course, stipulating that nudges cannot change financial incentives does not entail that nudged behavior is immune to economic consequences. Rather, the whole point of a properly calibrated nudge is to promote savings and avoid the undue costs that come from poorly designed behavior-modifying interfaces. The most accurate way to make this point, therefore, is to say that nudges are not pecuniary. With this point in mind, we can revisit an example discussed in Section 2, cafeterias that nudge consumers to eat less by shrinking the portions they serve. A more robust way of putting this point is to say that the cafeteria owners at issue must be motivated to make their consumers select healthy choices. If these owners shrink the portion size so that they can charge consumers more for the food they are serving, then the act of changing the plate size does not count as a nudge.

[6] Thaler and Sunstein also emphasize their vulnerability to cognitive bias to offset claims regarding the epistemic privilege of experts. See (Shrader-Frechette 2005).

developed a large literature on how values and politics influence the so-called "technical" assumptions and expertise that engineers employ when they design technologies and built environments (Winner 1980; Bijker et al. 1987; Law 1991; Bijker 1995; Franssen and Bucciarelli 2004; Feng and Feenberg 2008a,b; Vermaas and Pieter 2008). Paradigmatic examples include Langdon Winner's work on how race and class prejudices allegedly were built into the construction of bridges in Long Island, and Andrew Feenberg's analysis of *design space*, which clarifies how political context can constrain technical options. One of the aims of this research is to show how technical assumptions can be mediated by social and political values, and perhaps really are not "technical" at all (Bijker et al. 1987). Ultimately, critical theories like these aim to clarify the nuances of insufficiently understood social relations and political context. Unlike Thaler and Sunstein's project, they do not offer ways of changing people's behavior nor do they bracket politics. Despite this difference in orientation, it may be possible to bring the theory of choice architecture into productive dialogue with science and technology studies and the philosophy of technology. We will revisit this issue in Section 4 when we address the problem of meaning.

Perhaps, then, interface design is one of the easiest domains to see choice architecture at work. At present, considerable energy and resources are being devoted to projects that try to create *natural user interfaces*—interfaces that can perceive, communicate, and act smartly on our behalf by responding rapidly and intuitively to our bodily movements, gestures of touch, and acts of speech, and that will not play into biases we may have in those particular situations (Fogg and Brian 1997; Fogg 2003). In order to build these interfaces, engineers use tools associated with choice architecture, appropriating insights from behavioral economics and psychology, to anticipate how users will interpret and respond to various presentations of information. Of course, they also make use of other forms of behavioral and psychological knowledge. For example, choices people make about how to use their mobile phones are influenced by the size and position of the keyboard, the size and position of the screen, the geometry of the structures that link keyboard and screen, the software that governs how the phone functions and which present options for using different features in distinctly stylized ways, the phone's weight, and its size. Another relevant example is the graphical user interface tools and methods that allow users to control and manipulate computers and the applications that run on them. They include the keyboard and mouse, the acts of clicking and scrolling, and the presentation of menus, files and folders, et cetera. These examples of interface designs can be improved if their designers understand how to work with users' predictable biases.

In sum, nudges are a special kind of design calibration that is built for only one purpose: prompting better decisions by working with biases. The credit card console used for gaining entry to a parking garage is an example of choice architecture. The interface—which is constrained by the fact that there is only one possible direction for a credit card to be inserted—includes a diagram that serves as a heuristic for how the users should insert their cards (Thaler and Sunstein 2008, 89–90). There are many different diagrams that could be used for credit card consoles like this one. Whereas some diagrams will incline the majority of users to insert their card the wrong way, others will incline the majority get it right the first time. In cases where people insert their card the wrong way, they are likely relying on a bias developed from either previous experience with a console (or making bad inferences, as a result of having no prior experience). The diagrammatic interface plays a significant role in shaping decisions about how people decide to use the technology. This is no different from how, in cafeterias, the size of available plates influences how much food people will eat, and the arrangement of food options influences the items that hungry customers will select (Thaler and Sunstein 2008, 1–3). In each of these examples, nudging people is a matter of creating a better interface situation that will encourage them, whether they are aware of it or not, to make a better choice.

3. Libertarian Paternalism

Having just provided some additional details on nudges, we will now clarify why Thaler and Sunstein believe that nudges are permissible when they fall within the ethical limitations set forth by the principles of libertarian paternalism. We then argue, in Section 3, that even if libertarian paternalism is as attractive as *Nudge* stipulates, Thaler and Sunstein have not shown that choice architects who offer nudges merit our general trust.

Although libertarian paternalism may appear to be an oxymoron, Thaler and Sunstein defend it as an attractive moral outlook, which we will discuss as featuring three related principles. The first principle of libertarian paternalism is that *benefits and savings that improve lives are good*. The sorts of benefits and savings that Thaler and Sunstein refer to are those that individuals themselves view as such. They include common goods, such as increased health, improved safety, financial security, and so on. The second principle states that the *freedom to select one's own ends should be preserved*. In other words, the savings endorsed in the first principle should not be pursued through means that lead to other people determining our preferences and interests. Thaler and Sunstein only equate savings with benefits in cases where individuals themselves see their preferences and best interests as being served. In this respect, page after page of *Nudge* distances the text from hard paternalist depictions of people as poor judges of value in need of assistance from an interest-directing, benevolent *pater*. Furthermore, the second principle is proposed with the standard proviso that extreme situations exist where free choice should be limited, including emergencies, the avoidance of catastrophes, and cases of violent criminality. The third principle expresses the paternalistic side of libertarian paternalism. It states that *it is permissible and choiceworthy to help others achieve their self-directed ends when they cannot pursue these ends efficiently.* Situations where people do not have enough time or information to make deliberative choices are paradigmatic contexts where this last principle holds. If the other principles are not violated, Thaler and Sunstein state, then helping others in this way actually enhances peoples' ability to choose.

Thaler and Sunsteins' theory of nudges also includes a set of back-up principles for use in exceptional cases where compliance with the three core principles is insufficient. They insist that when conflicts of interest occur and when incentives cannot be lined up clearly, nudging is only permissible when the choice architect's design intentions are transparent and capable of being monitored (242). Nudges that cannot be made transparent and public thus are impermissible, as are ones that reflect racist, sexist, or other oppressive agendas. Such agendas could not be defended in the public sphere. Ultimately, Thaler and Sunstein deem the combination of back-up and core principles sufficient for demarcating nudges from exploitative behavior-modifying techniques, such as can be found in advertising and propaganda.

To illustrate the ethical limitations just discussed, let us consider briefly two examples of nudges that are consistent with libertarian paternalism. The first example brings us to Lake Shore Drive, a roadway that has stunning views of Chicago's skyline. One particular segment includes a series of S curves that require drivers to slow down to 25 mph. Many drivers ignore the posted

sign that states the reduced speed limit. They are easily distracted by the scenery, or else unable to assess how steep the curve is,
and both causes result in accidents. By introducing a new way of
nudging drivers to slow down, the individual and societal costs of
these accidents have been reduced. Immediately after passing a
warning sign, drivers encounter a series of white stripes that are
painted onto the road. Thaler and Sunstein describe this interface
as a prompt that inclines drivers to slow down:

> When the stripes first appear, they are evenly spaced,
> but as drivers reach the most dangerous portion of the
> curve, the stripes get closer together, giving the sen-
> sation that driving speed is increasing. One's natural
> instinct is to slow down. (38–39)

In short, the stripes work with drivers' tendencies better than
conventional signs because they convey the point about slowing
down intuitively and subtly. That is, the stripes do not require
drivers to interpret propositional information and think about how
they should behave in relation to considerations pertaining to the
abstract unit of miles per hour and a potentially arbitrary speeding scale. Rather, at an embodied level, they influence how *drivers
perceive* the turn, which becomes a way of decreasing the incid-
ence of bad decisions, thereby cutting the costs of accidents to
both individuals and other members of society.

The second example is Clocky, a special alarm clock. At some
point, we probably all have made plans to get up early on a given
morning in order to get a fresh start on the day. The plan seems
practical at night, but even when we have gotten enough sleep, it is
often hard to wake up because fatigue and the expected comfort
of additional sleep are too much to resist. Conventional alarm
clocks do not solve this problem for everyone. They are easily
turned off, or paused via hitting a snooze button. Clocky differs
from the rest, as it "runs away and hides if you don't get out of
bed" (Thaler and Sunstein 2008, 44). To use Clocky, one has to
set the acceptable number of snoozes and snooze minutes before
going to sleep. When all the snooze time is used up, the clock
literally springs off the nightstand and moves around the room
while making annoying sounds. The only way to turn it off is
to actually get out of bed and engage one's mental powers by
tracking it down. By the time Clocky is retrieved, the pursuer
can expect to be awake. Clocky's behavior helps people get up
when their will power and resolve require extra help.

Both the traffic stripes and Clocky examples count as nudges. In these cases, choice architects calibrate for the biases and blunders that constrain how some people cope with the relevant information. Crucially, neither example changes economic incentives, eliminates options, or makes freedom of choice difficult to exercise. After all, if one does not like Clocky, one need not use it. No one forces people to adopt it, and market competition is not skewed unfairly because the product exists. Likewise, the stripes simply yield an impression that a driver's speed might be dangerous for the upcoming curves. If one is determined, it is still possible to drive fast on the road and take in the spectacular views. With these examples in the background, Thaler and Sunstein inform their readers that libertarian paternalism functions as a viable ethical constraint that prevents nudges from being abused.

4. Trust and Competence

Though libertarian paternalism can be shown to provide certain ethical limitations on the possible uses of nudges, not everyone is persuaded by their account. Some critics claim that the foundations of libertarian paternalism are flawed, and even can be appealed to—at least in some cases—to endorse harmful outcomes (Stevenson 2005). While this is an interesting rejoinder, we will bracket assessment of it in order to focus on a related issue, *the competence required for a choice architect to offer nudges*. If our critique is valid, it will hold whether libertarian paternalism provides adequate normative constraints.

As Thaler and Sunstein define it, anyone can play the role of a choice architect, and do so in a variety of contexts. The appellation is not exclusive to engineers and designers, and Thaler and Sunstein ask everyone to offer nudges when in the position to do so. But even though most of us will encounter situations where it appears useful to offer nudges, it does not follow that everyone is *capable* of altering people's behavior appropriately. To this end, *Nudge* should clarify how choice architects can construct technologies, interfaces, and built environments that help to bring about desired and appropriately predictable outcomes in peoples' choices and behavior. Indeed, a convert to the nudge program who understands his or her role as a choice architect would still need to know how to prompt people subtly toward making better decisions when they do not have enough information or time, or are aroused and tempted in ways they had not anticipated.

Focusing on this issue leads us to ask: What kind of competence is required for a choice architect to offer a nudge? With this question in mind, we proceed by showing why Thaler and Sunstein's theory should include an account of competence. We emphasize the fact that since nudges typically change behavior without people being aware that they are being nudged, there ought to be reasons offered for why we should, *in general,* trust the competence of choice architects to design nudges that improve our lives.

It is implausible to believe that creating nudging choice architectures requires no competence whatsoever. According to our interpretation of Thaler and Sunstein, choice architects must be able to do two things *at a minimum.* First, they must be able to figure out what biases, arousals, and temptations people are subject to from studies in behavioral economics. Second, they must have an adequate understanding of how people perceive choice contexts. To do so, choice architects must have a sufficient grasp of the scientific material and a good understanding of how people think in particular situations. They must also be able to pick out the appropriate biases, arousals, and temptations that track people's thinking when they make choices in distinctive contexts and when presented with distinctive forms of information. Once choice architects identify the relevant mental stumbling blocks, they need to be able to postulate which calibrations in choice context will nudge people away from them. In other words, choice architects must grasp how people will perceive and respond to adjustments of their choice context. Without the ability to do so, there can be no basis for judging whether a nudge will succeed in altering people's behavior appropriately.

Unfortunately, Thaler and Sunstein do not discuss how choice architects are supposed to determine which biases, arousals, or temptations are relevant to a given situation, or how to arrive at appropriate proposals concerning the adjustment of choice context. This omission begs the question of whether such inferences and postulates can be made, especially since designing nudges requires that choice architects have the competence to make inferences from a limited body of empirical studies in behavioral economics and psychology. Additionally, choice architects have to possess an adequate understanding of the situations for which they intend to insert a nudge, and be able to come up with proposals about which nudges will encourage people to make better decisions, out of the host of possible adjustments that could be made to any situation. This is not a matter of whether choice architects can defend the purpose of a particular nudge; rather, it is a matter of whether choice architects can make the case that their nudge ideas will actually prompt people, on average, to make better choices.

It is unclear what sort of competence choice architects must have to be able to make these inferences and postulates. By asking for an account of competence, we have in mind an account of competence that would warrant our general trust in choice architects to offer nudges that would achieve the outcomes suggested by Thaler and Sunstein. By *general trust*, we mean the sort of trust that we ought to have of those whose testimony and products improve and influence our choices and behavior, such as scientists, engineers, lawyers, financial advisors, and the like. In the case of scientists, one of the characteristics which warrant our general trust in their testimony and research is their competence (Hardwig 1985, 1991). This is not only the case among scientists, but also between scientists and non-scientific members of the public (Scheman 2001; Rolin 2002; Wilholt 2009). For ordinary citizens to be able to benefit from the testimony and research of scientists, there ought to be reasons available to them to trust scientific testimony and research (Scheman 2001). Some of these reasons should be devoted to showing that scientists have the right kind of competence. In some ways, this is a matter of moral acceptability. Someone who wants to propose, for example, that science should play an increased role in some aspect of our lives, should be expected to show that the scientists in question are competent to do so. There is an analogy between this example of scientists and Thaler and Sunstein's nudges. Choice architects who offer nudges are producing changes in choice context that will allegedly improve and influence our choices and behavior. Because of this, the moral acceptability of nudges hinges, in part, on whether an account of competence can be provided that is sufficient to warrant our general trust in these products (the nudges).

To avoid misunderstanding, the emphasis that we are placing on competence does not entail that we believe that for every nudge offered, there should be good reasons for people to trust the competence of the individual choice architects who designed it. Rather, we are claiming that Thaler and Sunstein should be able to vouch for the competence of choice architects by offering general reasons for why their competence to offer nudges should be trusted. To be even more specific, we now will present four problems with competence and nudges that are rooted in Thaler and Sunsteins' not providing an account of competence.

(Problem of Inference) If choice architecture is essentially an idea based on select empirical studies of biases and anecdotal stories of bias correction, but remains detached from an adequate account of competence, then it is unclear how choice architects are supposed to use these studies and anecdotes as supporting evidence for nudges. Indeed, anyone could claim to be offering nudges based on ad hoc inferences underwritten by nothing more than the popularized account of empirical studies and peoples' perceptual habits that *Nudge* presents. Simply put, Thaler and Sunstein do not provide clear criteria for determining the minimal background conditions that need to be met in order for someone to be capable of claiming that they can offer a nudge based on appropriate consideration of the empirical studies. Here, we are not making substantive claims like choice architects should be able to understand technical papers in behavioral economics. Instead, we are playing the skeptic's role, and insisting that Thaler and Sunstein should clarify to what degree this is the case. This is reasonable because there are currently many efforts to understand how competence is related making inferences and judgments based on evidence, an example being the Studies in Expertise and Experience method at Cardiff University (Collins and Evans 2007).

This problem also includes the issue of how choice architects are supposed to get feedback on the successes and failures of nudges they have offered. If there is some competence associated with offering nudges, then it should be possible to identify reliable methods for obtaining feedback. This is particularly important in cases where nudges prove successful when first introduced, but fail to yield the intended results as time passes, perhaps as a result of users making new decisions upon learning how the choice architecture is configured (e.g., perhaps some users feel annoyed about being nudged, and subsequently challenge the behavior-modifying trajectory through defiant behavior that has the potential to catch on and inaugurate a counter program).

(Problem of Replication) If nudging is a good idea that should become more widely adopted, then the competence of choice architects should be defended against skeptics who may question whether choice architects can offer reliable nudges that fulfill their intended purpose. We may all agree that cases exist where people have been nudged. However, if there is no account of the techniques that willing choice architects should use to replicate successful nudges, then skeptics are entitled to claim that, given its basis in mere anecdotal evidence, choice architecture simply is not

the sort of endeavor that can be cultivated as a competence or expertise.[7] Skeptics could even claim that nudges are unacceptably risky to the people being nudged, inevitably leading to unintended consequences.

Another issue somewhat related to replication has to do with the sorts of available studies about successful nudges. Thaler and Sunstein insist that there is growing data confirming the success of certain nudges based on changing default settings in particular situations. But, if such confirming studies exist, the only justification lent by them is that the particular change in default setting produced the desired results. They do not establish that there is some general competence behind nudges that was used in the particular situation and that can be transferred to other situations— especially situations where it is not the default setting that requires change.

(*Problem of domains*) Thaler and Sunstein fail to clarify whether choice architecture is an independent science, technique, or expertise, or an adjunct to existing ones. If it is an adjunct, then the competence required to offer nudges largely depends on the competences and expertise associated with a professional domain. For example, if a group of choice architects are devising a plan to reduce speeding in a given area, then the competence at issue is the competence attributable to traffic calming professionals. However, if choice architecture should be conceptualized as an adjunct of this kind, then the following problem arises. Thaler and Sunstein do not specify how choice architecture can be integrated with the protocols, theoretical commitments, and tacit knowledge found in other domains. Nor do they clarify what exactly choice architecture is, such that it becomes possible to specify clearly what integration into another domain involves.

(*Problem of projection*) A fourth reason is raised by Mario Rizzo and Glen Whiman in "Little Brother is Watching," where they identify a variety of slippery slope possibilities that they believe justify skepticism towards nudges. Rizzo and Whiman insist that because choice architects necessarily have "only a tenuous grasp on the values of targeted agents," they can only concretely apply their paternalist ideas by making inferences about what the users of their designs are likely to value (Rizzo and Whiman 2008, 26).

---

[7] Some philosophers have taken on a project similar to Thaler and Sunsteins', but draw on a more complicated understanding of the sociality and materiality of design situations (Verbeek 2005).

In making such probabilistic inferences, Rizzo and Whiman assert, "there will be a tendency for the experts to reify their own values and simplify their own theories" (Rizzo and Whiman 2008, 26). When choice architects are unsure of the values that their targeted agents will possess, they will be inclined to perform the following four step process (Rizzo and Whiman 2008, 26–29):

1. Simplify the range of possible values by projecting their own contingently held predispositions onto a theoretical conception of genuine target audience preferences, such that if, for example, the experts possess "intellectual and middle class values," they will assume that the same values should obtain for everyone;

2. Justify using projection as the best means of accomplishing the needed simplification by associating the postulated ideals with "rational" thinking and the expected consequences of pursuing the identified "rational" ends with "optimal" outcomes;

3. Treat the representations of expected preferences as isomorphic depictions of what the targeted agents definitively desire, and not fictions that were selected for pragmatic reasons;

4. Obscure their "ethical" decision to use projection as a means of simplification by acting as if objective scientific principles were used to bridge the knowledge gap when, in fact, "neither scientific theory nor scientific evidence provided the basis for favoring one preference ordering over another."

Rizzo and Whiman's concerns are similar to what we termed postulates earlier in this section. Without an account of how choice architects can make competent postulates or projections, we have no reason to expect choice architects to make good inferences about what people's preferences are. Thus, nudges are only good ideas if we can be sure that we can reliably know people's preferences. But, how can we know this?

It might be claimed that the four reasons just offered are really not good reasons at all because, ultimately, the competence of choice architects and the success of nudges is a function of *trial and error*. This counter claim, however, does not match up with current controversies over how to design interfaces, which demonstrate that, especially for innovative technologies that promise public benefits, reasons need to be provided in advance for why a particular methodology, like choice architecture, should be trusted over its rivals.

Consider recent NPR broadcast on smart meters that featured Dan Reicher, Director of Climate Change and Energy Initiatives at Google Inc., and a Carnegie-Mellon behavioral economist George Loewenstein. In it, Reicher took the more *information* is always better stance, which presumes that people will make better decisions when they have access to all salient information. Loewenstein, to the contrary, suggested that the informational interface provided by smart meters may actually activate biases that incline people to make costly decisions:

> It's amazingly cheap to air-condition your whole house for a few hours. And if the smart meter is giving you objective information about how much it's costing you, you might be surprised at how cheap it is rather than surprised at how expensive it is." Accordingly, Lowenstein suggested more automation in some aspects that people could not control based on biased judgments of the cost of energy. Debates like this one impede the ability to set up trial and error tests precisely because they convey disagreement over what type of interface should be tested. Indeed, the dispute at issue concerns different hypotheses about how people perceive smart meter interfaces, and different judgments about which, biases—if any— can impede users from using smart meters efficiently. Reicher assumes that it is possible to match information displays to people's lifestyles in ways that allow them to make free yet more efficient choices about how much energy to use. By contrast, Lowenstein assumes that certain biases will make it more likely that people will make more costly decisions. Without reasons available to defend which side is correct, we have no reason to favor the plausibility of either side, or to expect that they exhaust the spectrum of possible interface designs. This problem extends to *Nudge* where Thaler and Sunstein suggest that an ambient orb can nudge users towards energy efficient behavior by distinctively conveying how much energy a household is using. (Thaler and Sunstein 2008, 196)

So far, we have shown that the absence of an account of competence in *Nudge* has implications that compromise Sunstein and Thaler's defense of the moral acceptability of choice architecture. But, are there good reasons that *could* be offered for why we should trust choice architects to offer nudges? If so, these reasons would, in effect, have to allay the concerns just raised about competence. In Section 4, we extend the discussion beyond the general methodological issues raised here and argue that if such reasons were to be provided, then there is a particularly hard problem that such reasons would also have to resolve. While Rizzo and Whiman focus on choice architects' postulates about preferences, we believe that choice architects have to be competent at postulating how any calibrations will affect people's perception of the *meaning of those calibrations*, especially since the meaning is likely to change in ways that are difficult to predict. We refer to this as the problem of *semantic variance*, which we will describe in more detail in the next section. A good account of competence should track semantic variance and provide reasons for why choice architects are able to design nudges that are sensitive to it.

5. Semantic Variance

As a basic definition that serves the purpose of this paper, meaning refers to significance. An invitation to smoke a Cuban cigar can mean different things to different people because each person can perceive the invitation as having different significance. For example, an aesthete can perceive the invitation as an opportunity to enjoy a pleasurable experience. A US citizen can perceive the invitation as an opportunity to enjoy a risky experience with contraband material. A Cuban expatriate can perceive the invitation as an opportunity to have a nostalgic experience of home. Many other possibilities exist; this brief list is being used for the sole purpose of concretizing a definition.

Meaning is a crucial feature of Thaler and Sunsteins' account of nudges because they write in a way that presupposes choice architects can competently identify people's perceptions of meaning. An illustrative example of this presupposition can be found in their discussion of how fly-etched urinals in the men's rooms in Amsterdam's Schiphol airport significantly reduces spillage.

Small and apparently insignificant details can have major impacts on people's behavior. A good rule of thumb is to assume that "everything matters." In many cases, the power of these details comes from focusing the attention of users in a particular direction. A wonderful example of this principle comes from, of all places, the men's rooms at Schiphol Airport in Amsterdam.

There the authorities have etched the image of a black housefly into each urinal. It seems that men usually do not pay much attention to where they aim, which can create a bit of a mess, but if they see a target, attention and therefore accuracy are much increased (Thaler and Sunstein 2008, 3–4).

While Thaler and Sunstein implore choice architects to consider "everything," their own attention is focused selectively. In this example, only two variables are identified as being worthy of consideration: the normal attention men exhibit when urinating in public restrooms, and the capacity of a target to capture their attention in this context. While these considerations are sensible, they do not account for all relevant possibilities.

For example, Thaler and Sunstein do not ask if it is a universal fact that under certain conditions men exhibited shortened attention span, or if only some do, and perhaps for culturally or personally contingent reasons. Nor do they inquire into whether certain predictable bodily responses to targets are more likely to occur in the case of bodies that have been culturally disciplined to behave in distinctive ways. Furthermore, Thaler and Sunstein do not examine whether the black housefly, which is the core choice architecture contribution that changes the standard urinal interface, is a value-laden symbol. As a thought experiment, we can imagine a culture existing that exhibits such deep reverence for all life that its members would be offended by the prospect of someone urinating on a representation of an insect. Their outrage could be extreme, and parallel the indignation that select Muslim communities felt over the infamous Danish political cartoon of the prophet Muhammad!

We would remind the reader who only sees our thought experiment as an unrealistic depiction of a far-fetched cultural reaction, that the editorial staff of the Danish newspaper *Jyllands-Posten* did not anticipate that publishing the cartoons would lead to outspoken denunciations, legal motions alleging that violations of the Danish Penal Code occurred, consumer boycotts, acts of retaliatory violence being committed against Danish embassies, death threats being made against those responsible for the cartoons, as well as a host of other unpleasant consequences. But even if it is scarcely conceivable that a culture could exist that would pose death threats to people who design fly-etched urinals, the point of principle illustrated by the thought experiment merits further consideration. Thaler and Sunstein may have made a lucky pick when selecting a nudge that contains only innocuous detailing. Alternatively, they may have rhetorically disguised a heavy-handed example by making it appear to be an interface that choice architects could design solely by applying behavioral economics insights into cognitive bias. Only these two possibilities exist, as Thaler and Sunsteins' account of nudges presupposes that choice architects know how to calibrate choice contexts to capitalize on commonly shared perceptions of meaning, even though they never ask the following two questions that should be considered basic to any design initiative that aspires to shape people's behavior responsibly.

- How do technologies, interfaces, and built environments come to be invested with meaning?

- How does the meaning attributed to a technology, interface, and built environment change?

Both questions refer to the *problem of meaning*, a perceptual, epistemological, and sometimes *political* issue investigated by scholars in philosophy of technology and science and technology studies who analyze the significance attributed to material culture.

We believe that Thaler and Sunstein underestimate how difficult it can be to understand and predict how different communities of people will perceive the meanings that nudges present because *Nudge's* illustrative examples all focus on situations where technologies, interfaces, and built environments are used in delimited contexts. In these contexts, (1) users appear to have common perceptions of meaning and (2) user interaction with the technologies, interfaces, and built environments does not appear to engender new perceptions of meaning. These contexts can be uniformly characterized as instances of *semantic invariance*, and a brief comparison with a relevant example discussed by science and technology studies theorist Bruno Latour will illuminate some of its features.

In a frequently cited passage, Latour analyzes a typical key found in European hotels that is bound by a cumbersome weight (Latour 2000, 41). The weight was added to the key to solve the problem of guests failing to return their keys to the concierge or hotel manager before leaving the hotel. Latour notes that the cumbersome weight is a more effective behavioral prompt than some of the other discursive solutions, such as posted reminders and inscribed keys (Latour 2000).

In this case, the hotel management clearly nudges the guests to return their keys before leaving. In many respects, they offer a key that exhibits the same design principles as any generic key. Its difference lies in the special interface, one that binds the key to a cumbersome weight and therein changes the arrangement of user options, nudging guests toward a distinctive outcome without altering any of the relevant incentive structures. Crucially, the more guests return their keys, the less the hotel incurs the costs associated with their mismanagement. Since such costs are often passed on to customers through penalty fees, savings are enhanced all around.

Designers can be choice architects in this instance because whether weighted or free, guests and managers can be expected to perceive the hotel keys as conveying two meaningful ideals: access and security. Indeed, although the key's interface has been altered, nothing has been done to modify the guest's perception that a key's main purpose is to open and lock doors. In that situation, the key's design can be adjusted without affecting anything else but the decision of the guest to avoid wandering around with the key in his or her possession. All things being equal (e.g., the quality of the stay, the cost of a room, the lack of desire for an illicit souvenir, and so on), adding the cumbersome weight simply makes it easier for guests to come to a non-controversial and self-benefitting decision.

Thaler and Sunstein's examples of Clocky and Lake Shore Drive have a similar *semantically invariant* structure. Adding the nudge of the loud, annoying, and hiding routine, changes nothing else other than supporting the person's decision to wake up at the intended time; adding stripes to the road helps people to slow down, but does not engender any other significant changes. In these examples, the nudges appear to be effective because people perceive common meanings, and no new perceptions of meaning are generated.

It is problematic to construct a theory of choice architecture exclusively around examples of *semantic invariance* because such examples do not capture the full range of situations where people interact with technologies, interfaces, and built environments. In many cases, such interactions occur under conditions of *semantic variance*, which means that diverse perceptions of meaning occur. To concretize this point, let us consider the raised surface of a speed bump, which provides a disincentive for people to pass over it too quickly, as approaching drivers readily recognize that the artifact can damage their cars and provide an uncomfortable, bumpy experience. Latour provocatively characterizes speed bumps as "actants" and reminds us that they have been referred to as "sleeping policemen" by virtue of their capacity to perform the same functional role as law enforcement officers. Using a similar, albeit more prosaically expressed perspective, Thaler and Sunstein characterize "make-believe speed bumps," which are "painted 3-D triangles that look like speed bumps" but cost much less to make than the real ones, as nudges (Thaler and Sunstein 2008, 261).

The problem, here, is that categorically depicting speed bumps as nudges begs the question of how perceptions of meaning can vary. In some contexts, such as speed bumps being placed in roads adjacent to schools, most people likely will see the artifact in just this way. However, as science and technology studies theorist Trevor Pinch clarifies in "On making infrastructure visible: putting the non-humans to rights," several contexts exist in which proposals concerning the use of "traffic calming devices" ranging from speed bumps to cobbled shoulders leads to acrimonious debate; diverse users attribute different meanings to the artifacts (Pinch 2009). Partisans who champion the cause of maintaining infrastructure that is supportive of safe bicycling can readily clash with partisans who place a higher premium on efficient automobile travel or pedestrian rights. While such debates ostensibly concern the desirability of given traffic calming proposal, the contested values underlying the debates are broader, often involving issues related to environmental sustainability, economic expense, life-enhancing aesthetics, and the difficulties that attend to allowing old technologies (e.g., bicycles), to be used under material conditions that are designed to support new ones (e.g., cars; Pinch 2009).

To clarify further why *semantic variance* poses a potential problem for Thaler and Sunstein's needed account of competence, we will now offer brief discussion of the following examples: (1) a standard global positioning systems (GPS) designed for car use and (2) a program that uses a mobile phone's photographic capacities to help people eat better, (3) an exercise promoting program on Nintendo Wii and (4) a proposal for increasing organ donations, and (5) the Google buzz program for Gmail. Each of these examples is similar to examples of nudges provided in Nudge (and covered previously in this paper), and (4) is an actual example of a nudge detailed in the book. We will show how each example suggests that perception of meaning may change depending on the *contexts* and the *identities* of the people involved in the situation.

Though context and identity issues permeate each example, for analytic reasons, we will tend to emphasize one or the other for each example.

1. GPS devices are designed to make it easy to navigate from one destination to another by providing drivers with step-by-step prompts (e.g., turn left or right) that guide a trip from start to finish. Such prompts are especially useful given the information processing limits of the typical human mind. Recently, GPS systems have been designed to do more than help drivers cope with the natural limits. They now nudge drivers away from speeding. In the case of a popular TomTom model, once one starts breaking the speed limit, a notification comes on the screen that is highlighted in red, a color that evokes stop signs and the stop signal of a traffic light. The driver is then able to see, at a glance, the difference between the speed he or she is travelling at, and the speed that is legal to be driving at. The perception of this disparity is intended to motivate drivers to slow down. However, one of the authors of this paper experiences the TomTom as having precisely the opposite effect! When he notices that he is speeding, he also notices that the information listing how much time remains before the trip is completed becomes shortened. Seeing the reduced trip time changes the meaning of speeding in his perception and experience. Rather than that awareness triggering a desire to minimize the likelihood of causing an accident or getting a speeding ticket, it actually prompts him to try to reduce the trip time by visually measurable increments—5–30 min, depending on the length of the trip—that correlate with affectively charged responses. As if playing a videogame, the driver finds himself increasingly satisfied as those increments go down.

   People do usually speed because they are not paying attention to the odometer, not reflecting on the consequences of speeding, and focusing on phenomenological features that do allow them to register just how fast they are going. It thus would seem perfectly reasonable to use these assumptions to build-in the feature just described so as to decrease the amount of drivers speeding on the road. But, as our example suggests, the choice architect is going to have to understand how the *contexts* in which TomToms are used will change how the meaning of the notification is perceived by drivers. The choice architect will have to have a sufficient grasp on how various contexts relate to the perception of meaning and how a notification in general will reduce speeding across these contexts.

2. The FoodPhone program allows mobile phone users to take pictures of food they plan on eating and electronically send the images to dietary experts who promptly respond with nutrition guidance, the intended result being that people will naturally make better food choices when they lack time and information to find out for themselves. Dutch philosopher of technology Peter-Paul Verbeek, however, depicts the program as a potential conduit of the following two externalities, all of which relate to the changes in meaning that the technological adjustment can engender when it mediates the experience of the person using it. First, it can make the activity of eating unduly stressful by transforming meal time into a period of constant judgment. Second, it can incline participants to view health through the overly narrow lens of food consumption and it can negatively impact the quality of the social relations that transpire around shared meals by nudging participants to obsess over their food and act more like observers of eating behavior than absorbed participants in a communal experience (Verbeek 2009). Like with the TomTom, the context in which the FoodPhone is used may change how people perceive the meaning of the situation. Whereas the FoodPhone was intended as a subtle prompt, it may actually make people overly reflective and stressed out, overtly aware that they do not have enough time and information and yet must attend to every bite. Can choice architects account for how perception of the meaning of choice context changes with context?

3. The videogame Wii Fit is marketed as an entertainment system that can help players of all ages enhance their fitness through fun exercises. To inspire users to stay on track of their fitness goals, the Wii Fit scale makes groaning sounds when players gain weight, and it also analyzes players' body-mass-index, providing them with correlative qualitative labels, such as underweight, ideal, and fat. The problem, though, is that different users can attribute different meanings to these outputs. Controversy thus arose when young girls, a population that is especially vulnerable to concerns

related to body image, were informed they were *fat*. Their parents perceived this as demeaning and complained that it is myopic to view the interface solely as inspiring healthy living. A similar outcome occurred when one of the authors tried playing sports games on Wii Fit with his young daughter. These games depict successful performance through avatars that express positive body language and unsuccessful performances through avatars that look downtrodden. While these outputs might nudge adults to do better, it had the opposite effect on his daughter. She was very upset to find her avatar looking despondent. Moreover, because she perceived such strong meaning in the body language conveyed, she refused to believe the white lies her concerned parents offered to try to make the situation less frustrating (i.e., that the avatar was tired, not sad). In this case, some of the features of Wii Fit are overt and some are subtle and similar to nudges. Here, we do not want to highlight context in relation to meaning, but *identity*. The subtle prompts toward weight loss and persistent game play will be perceived differently according to the identity of person using the technology. Choice architects would have to be able to have some understanding of the relation between identity and meaning simply to avoid unintended harms like hurt feelings and low self-esteem.

4. In order to increase the rate of organ donation in the US, Thaler and Sunstein suggest it is worth considering instituting a new default setting called "presumed consent" (Thaler and Sunstein 2008, 179). Unlike "explicit consent," which requires that citizens take active steps to demonstrate that they want to be organ donors, presumed consent would be guided by the ideal that "all citizens" should be "presumed to be consenting organ donors," unless, through some easily available means, like checking an opt-out box when applying for a driver's license, they specify otherwise (Thaler and Sunstein 2008, 181–182). While they acknowledge that such a nudge passes the libertarian paternalism test, they also concede that to avoid making citizens unduly upset about such a "sensitive matter," it might be best to pursue the less radical option of "mandated choice"—an option that simply requires that everyone who applies for a driver's license explicitly check off a box that indicates their preference to donate or not donate organs upon death.

Bioethicist Art Caplan correctly points out that even this less radical proposal fails to be viable precisely because Thaler and Sunstein have not raised the appropriate questions related to meaning and identity.[8] According to Caplan, what Thaler and Sunstein do not appreciate is the fact that in the US, many people are so skeptical about the motivations guiding a range of healthcare workers that mandated choice option would backfire and lead to a decrease in organ donations. Caplan speculates that people would be too afraid to check off the box out of concern that doing so would provide incentives for healthcare workers to provide them with bad treatment in order to obtain organs.

5. Google Buzz is a social networking service that allows Gmail users to see the contents of other peoples Gmail accounts, similar to Facebook and Twitter. This includes photos, videos, status updates, and more. The people at Google assumed that users should be automatically opted-in to Google buzz when they sign up for Gmail. This would prompt people to sign up for the program and receive its benefits to them as a default option. The problem that arose—and which is indicated by several class action suits—is that many people perceive being automatically opted-in to Google buzz as a violation of privacy, and that such a default option, even if it would benefit many people who normally would not have enrolled in Google buzz, would not be considered beneficial to everyone. In fact, it appears as if Google made the decision to automatically enroll without any competent evaluation of the data or information on Gmail users, which left open the possibility that many people would be upset. The particular context of Google Buzz and the identities of Gmail users had a lot to do with how the meaning of the default option was perceived, and ultimately created difficulties for Google.

Each of these examples illustrates simple semantic variations that can occur when the choice context is calibrated. Sometimes semantic variations can lead to harms, as in the case if Wii Fit; in other cases, the intentions of the designer are undermined for some people, but perhaps not for others. The point is that choice situ-

---

[8] Personal correspondence.

ations include multiple ways in which the meaning of the choice
context is perceived by those who inhabit them. By appealing to
these examples, we are not making the point that semantic variance will always occur in some sense that rules out Thaler and
Sunstein's approach in *Nudge*. We include these examples only to
suggest that semantic variance is something for which Thaler and
Sunstein should account if they are to convince of us of the moral
acceptability of nudges. Our examples only allow us to make the
claim that semantic variance is a significant concern, and that
any reasons for why choice architects who offer nudges should be
trusted would have to cover how they can negotiate semantic variance. If there are reasons why choice architects should be trusted,
at least some of these reasons have to be able to show how choice
architects can competently anticipate semantic variance.

6. Conclusion

Our discussion of semantic variance draws on science and technology studies and the philosophy of technology. Because Thaler
and Sunstein restrict *Nudge's* examples to cases of semantic invariance, philosopher of technology Don Ihde characterizes their
text in the following negative terms:

> While Thaler and Sunstein are indeed more inventive
> and original than the econometric and technocratic
> pack that they run with, they remain through and
> through econometric and technocratic. That is, the
> people who inhabit their world are cardboard charac-
> ters..These guys, like the Cold Warriors before them,
> think in a weird world—not one that I'd call a Life-
> world. It's more a world inhabited by Heideggerian,
> calculating robots than messy emotional humans. Their
> world is like the one Heidegger fears, now already as-
> sumed to be the real one.[9]

Ihde's point is well-taken, albeit only as an assessment of how
Nudge is written, and not a decisive judgment about whether
Thaler and Sunstein in principle can provide an account of competence that would respond to the general methodological issues
raised in Section 3 and the problem of tracking *semantic variance*
discussed in Section 4. Whether such an account could be given,

---

[9] Personal correspondence.

and done in a way that makes the proper connections between trust and competence, is an open question, but one that many science and technology studies scholars and philosophers of technology would be skeptical of given how social and material reality is framed in *Nudge*

We hope that the present essay succeeds in furthering the conversation about the nature and scope of choice architecture, and helps clarify fundamental issues that choice architecture proponents like Thaler and Sunstein need to address. Although our critical remarks have focused on omissions in *Nudge*, our structuring emphasis on the relation between competence and trust should be understood as falling directly in line with the type of project for which Thaler and Sunstein advocate. While examples like Clocky are interesting, Thaler and Sunstein appear to seek more than a set of interpretations of these examples, but an useful strategy for increasing savings, cutting costs, and getting people more of what they consider to be valuable by their own lights. This alternative matters, according to Thaler and Sunstein, because nudge projects can be high stakes endeavors. For example, if switching the presentation of a human resource document for employee savings did not actually lead to enhanced savings, then employers and employees alike would have good reason for feeling disappointed, and perhaps even betrayed.

We conclude by suggesting that choice architects can grasp semantic variance if Thaler and Sunsteins' approach to design is compatible with insights about meaning expressed in science and technology studies and philosophy of technology. Further multidisciplinary and collaborative research should be undertaken that emphasizes the qualities of various approaches to understanding how people make decisions when they interface with technologies, artifacts, and built environments. Perhaps there are important projects ahead that unite behavioral economics, science and technology studies, and the philosophy of technology.

Bibliography

Baier, Annette. (1986). Trust and anti-trust. *Ethics* 96: 231-260.

Bijker, Wiebe E. (1995). *Of bicycles, bakelites, and bulbs: Toward a theory of sociotechnical change, Inside technology.* Cambridge, Mass.: MIT Press.

Bijker, Wiebe E., Thomas Parke Hughes, and T. J. Pinch. (1987). *The social construction of technological systems: New directions in the sociology and history of technology.* Cambridge, MA: MIT

Press.

Cerbone, David. (2009). (book 1) ironic technics; (book 2) post-phenomenology and technoscience: The peking university lectures. *Notre Dame Philosophical Reviews.*

Collins, H. M., and Robert Evans. (2007). *Rethinking expertise.* Chicago: University of Chicago Press.

Epstein, S. (1994). Integration of the cognitive and the psychodynamic unconscious. *American Psychologist* 49: 709-724.

Feng, Patrick, and Andrew Feenberg. (2008). Critical theory of technology and the design process. In *Philosophy and design,* edited by P. E. Vermaas, Kroess, P.A., Light, Andrew, and Moore, Steven. New York, NY: Springer.

(2008). Thinking about design: Critical theory of technology and the design process. In *Philosophy and design: From engineering to architecture,* edited by P. E. Vermass, P. Kroes, A. Light and S. A. Moore. Dordrecht, NL: Springer.

Fogg, Brian J. (1997). Charismatic computers: Creating more likable and persuasive interactive technologies by leveraging principles from social psychology. Thesis (Ph. D.), Stanford University.

(2003). *Persuasive technology: Using computers to change what we think and do.* Amsterdam ; Boston: Morgan Kaufmann Publishers.

Franssen, Maarten, and Louis L. Bucciarelli. (2004). On rationality in engineering design. *Journal of Mechanical Design* 126 (6): 945-949.

Hardwig, John. (1985). Epistemic dependence. *Journal of Philosophy* 82 (7): 335-349.

———. (1991). The role of trust in knowledge. *Journal of Philosophy* 88 (12): 693-708.

Hieronymi, Pamela. (2008). The reasons of trust. *Australasian Journal of Philosophy* 86 (2): 213 – 236.

Hinchman, Edward. (2005). Advising as inviting to trust. *Canadian Journal of Philosophy* 35 (3): 355-386.

———. (2005). Telling as inviting to trust. *Philosophy and Phenomenological Research* 70 (3): 562–587.

Holton, Richard. (1994). Deciding to trust, coming to believe. *Australasian Journal of Philosophy* 72 (1): 63-76.

Ihde, Don. (1977). *Experimental phenomenology: An introduction.* New York: Putnam.

————. (2007). *Listening and voice: Phenomenologies of sound.* 2nd ed. Albany: State University of New York Press.

Jones, Karen. (1996). Trust as an affective attitude. *Ethics* 107 (1): 4-25.

Latour, Bruno. (2000). Technology is society made durable. In *Work and society: A reader*, edited by K. Grint. Cambridge: Polity.

Law, John. (1991). *The sociology of monsters: Essays on power, technology and domination.* London: Routledge.

Levitt, Steven D., and Stephen J. Dubner. (2005). *Freakonomics: A rogue economist explores the hidden side of everything.* 1st ed. New York: William Morrow.

Lobel, Orly, and On Amir. (2009). Stumble, predict, nudge: How behavioral economics informs law and policy. *Columbia Law Review* 108: 2098-2139.

Pinch, Trevor. (2009). On making infrastructure visible: Putting the non-humans to rights. *Cambridge Journal of Economics* 34 (1): 77-89.

Rizzo, Mario J., and Douglas G. Whiman. (2008). Little brother is watching you: New paternalism on the slippery slopes. *SSRN eLibrary.*

Rolin, Kristina. (2002). Gender and trust in science. *Hypatia: A Journal of Feminist Philosophy* 17 (4): 95-118.

Scheman, Naomi. (2001). Epistemology resuscitated: Objectivity as trustworthiness. In *Engendering rationalities*, edited by N. Tuana and S. Morgen. Albany: SUNY Press.

Selinger, Evan. (2006). *Postphenomenology: A critical companion to Ihde.* Albany: State University of New York Press.

Shrader-Frechette, Kristin. (2005). Flawed attacks on contemporary human rights: Laudan, Sunstein, and the cost-benefit state. *Human Rights Review* 7 (1): 92-110.

Stevenson, Drury. (2005). Libertarian paternalism: The cocaine vaccine as a test case for the Sunstein/Thaler model. *Rutgers Journal of Law and Urban Policy* 3:

Taddeo, Mariarosaria. (2009). Defining trust and e-trust: Old theories and new problems. *International Journal of Technology and Human Interaction (IJTHI) Official Publication of the Information Resources Management Association* 5 (2): 23-35.

Thaler, Richard, and Cass Sunstein. (2003). Libertarian paternalism. *The American Economic Review* 93 (2): 175-179.

————(2003). Libertarian paternalism is not an oxymoron. *University of Chicago Law Review* 70 (4): 1159-1202.

————(2008). *Nudge: Improving decisions about health, wealth, and happiness.* New Haven: Yale University Press.

Verbeek, Peter-Paul. (2005). Artifacts and attachment: A postscript philosophy of mediation. In *Inside the politics of technology,* edited by H. Harbers. Amsterdam: University of Amsterdam Press.

————. (2009). Ambient intelligence and persuasive technology: The blurring boundaries between human and technology. *NanoEthics* 3 (3): 231-242.

Vermaas, Pieter E. (2008). Philosophy and design from engineering to architecture. [Dordrecht]: Springer. http://rave.ohiolink.edu/ebooks/ebc/9781402065910.

Wilholt, Torsten. (2009). Bias and values in scientific research. *Studies in History and Philosophy of Science* 40: 92-101.

Winner, Langdon. (1980). Do artifacts have politics? *Daedalus* 109 (1).

Wright, Stephen. 2009. Trust and trustworthiness. *Philosophia* 38 (3): 615-627.

8
About the Author

Evan Selinger is Associate Professor of Philosophy and Graduate Program Faculty Member in the Golisano Institute for Sustainability, both at Rochester Institute of Technology. He has published extensively in the areas of philosophy of technology, ethics and policy of science and technology, phenomenology, and applied ethics. His latest co-edited books are *5 Questions: Sustainability Ethics* (Automatic/VIP Press 2010), *Rethinking Theories and Practices of Imaging* (Palgrave McMillan 2009), and *New Waves in Philosophy of Technology* (Palgrave McMillan 2009). He is currently Executive Editor of the journal *Philosophy and Technology* (Springer).

Index

www.ingramcontent.com/pod-product-compliance
Lightning Source LLC
Chambersburg PA
CBHW021556210326
41599CB00010B/472